Coding and Quantitative Biology Series

R Crash Course for Biologists
R STATS Crash Course for Biologists
R Machine Learning Crash Course for Biologists

For latest titles and links, see:
https://EcoEvoGeno.org

R Crash Course for Biologists

An introduction to R for bioinformatics and biostatistics

Robert I. Colautti

Biology Department
Queen's University (Canada)

Dragon Phylogeny Press • Kingston, ON CANADA

Dragon Phylogeny Press
Kingston, ON CANADA
https://DragonPhylogeny.org

R Crash Course for Biologists
ISBN: 9798849115917
Independently published by the author

By Robert I. Colautti
Queen's University,
Kingston, ON K7L 3N6 Canada

Copyright © 2022 Robert I. Colautti. All rights reserved.

November 2022: First Edition. Revised: January 2023.

The author has operated in good faith to ensure this work is accurate, but as a self-published book there can be no guarantee. It should go without saying that the author does not claim responsibility for any errors or misuse of the information contained herein. But let's be honest, you aren't reading this unless you are publishing your own book and want to see what other people put here.

> For the Yakilautti family

[1] (both biological and honorary)
[2] and for the survivors
[3] of R Crash Course for Biologists

Contents

Preface   **9**

    0.1   Is this book for you? 10

    0.2   Why R? 12

    0.3   Advice 13

    0.4   Learn By Doing! 16

    0.5   What to Expect 17

    0.6   Translational Coding 18

Setup   **23**

    0.7   R 23

    0.8   R Studio 23

    0.9   Console 24

    0.10 Packages 25

1 R Fundamentals   **27**

    1.1   Overview 27

    1.2   R Basics 28

1.3	Use ? for HELP	37
1.4	Random Numbers	39
1.5	Repeat, Replicate & Sample	42
1.6	`set.seed()`	44
1.7	Combining objects	46
1.8	Sequence	47
1.9	Rows and Columns	48
1.10	Data Types	50
1.11	Objects & Variables	52
1.12	Matrix Algebra	60
1.13	PCA	67
1.14	Lists	69
1.15	`print()` and `paste()`	73
1.16	External Files	75
1.17	Other Functions	86
1.18	Tidyverse	90
1.19	Save	91
1.20	Packages	92
1.21	Readable code	95

2 Flow Control — 97

2.1	Overview	97
2.2	`if(){}`	98

	2.3 `ifelse()`	98
	2.4 nested `if`	99
	2.5 for loop	100
	2.6 Nested Loops	103
	2.7 while loop	104
	2.8 Modulo	106
	2.9 Faster loops	107
3	**Quick Visualizations**	**111**
	3.1 Overview	111
	3.2 Graphical Concepts	113
	3.3 Getting Started	119
	3.4 Basic Graphs	122
4	**Basic Customization**	**133**
	4.1 Overview	133
	4.2 Setup	133
	4.3 `binwidth`	134
	4.4 `size`	135
	4.5 `alpha`	138
	4.6 `colour` (or `color`)	138
	4.7 Colour with `rgb()`	140
	4.8 `fill`	143
	4.9 `position`	144

- 4.10 shape 144
- 4.11 `lab`, `xlab`, and `ylab` 147
- 4.12 `labs` 148
- 4.13 Themes and Geoms 149
- 4.14 Basic Multi-Plot Graphs 156
- 4.15 Graph output 158
- 4.16 Practice 160

5 Advanced Visualizations 161
- 5.1 Overview 161
- 5.2 Getting Started 162
- 5.3 Rules of thumb 162
- 5.4 Example 169
- 5.5 Measuring Selection 173
- 5.6 Distribution Plots 176
- 5.7 Full ggplot 184
- 5.8 Bootstrap 185
- 5.9 Plot data 187

6 Multi-plot Graphs 193
- 6.1 Overview 193
- 6.2 Setup 193
- 6.3 `facets` 194
- 6.4 `gridExtra` package 196

       6.5 `grid` package 198

       6.6 Further Reading 201

7 **Regular Expressions**   **203**

       7.1 Overview 203

       7.2 Functions 205

       7.3 Wildcards 208

       7.4 New Lines 212

       7.5 Special characters 213

       7.6 [ ] (set) 219

       7.7 ^ (start or negate) and $ (end) 220

       7.8 ( ) (capture) 221

       7.9 Scraping 222

       7.10 Examples 225

       7.11 Transect Data 225

       7.12 Genbank 226

       7.13 Solutions 229

       7.14 More Exercises 231

8 **Data Science**   **233**

       8.1 Overview 233

       8.2 Setup 235

       8.3 2D Data Wrangling 236

       8.4 Join datasets 252

- 8.5 *Wide* vs *Long* data 255
- 8.6 Missing Data 257
- 8.7 Naughty Data 261
- 8.8 Dates 263

9 R Markdown 273

- 9.1 Overview 273
- 9.2 Setup 274
- 9.3 Cheat Sheet 274
- 9.4 Create 274
- 9.5 YAML Header 276
- 9.6 Markdown Elements 276
- 9.7 Basic elements: 277
- 9.8 Headers 278
- 9.9 Other Elements 278
- 9.10 Links 278
- 9.11 Images 279
- 9.12 Lists 280
- 9.13 Tables 282
- 9.14 Tables 282
- 9.15 Embed R Code 283
- 9.16 Dynamic tables 285
- 9.17 Embed Graphs 286

9.18 Content as tabs 287

9.19 Equations 289

10 Custom Functions   293

10.1 Overview 293

10.2 General form: 294

10.3 Example function 295

10.4 Local vs Global 296

10.5 Run custom functions 297

10.6 Annotation 299

10.7 Verbose parameter 301

10.8 External files 302

11 Conclusion   305

11.1 You made it! 305

11.2 What next? 308

11.3 Support Open & Accessible Science 310

11.4 Picture a Coder 311

Preface

Think of a Biologist. Who do you see? Take a minute to write down some characteristics in your mind. Try to be specific: gender, skin, age, height, hair, clothes, personality. Who do you see?

Now think of a *computer programmer* or *data scientist*. Write down their characteristics. How do these people differ in your mind? Can you imagine them being the same person? Can you picture yourself in both roles?

The goal of this book is to bridge these two worlds. In writing this book, I assume you are a practising biologist or a student of biology, or you are just motivated by biological phenomena. It doesn't matter if you are a recent high school graduate entering into a biology undergraduate program, a graduate student embarking on an independent research dissertation, or a senior scientist with specialized expertise in the science of life. As long as you are interested in learning how to code, this book is written for you.

The goal of this book is to provide a 'how-to' guide to connect you to the world of data science. We focus on the fundamentals of the R programming language and its applications in biology. In writing this book, I assume you do not have much coding experience. Whether you are a new biology student or a seasoned professional, this book was written for you.

There are many great introductions to the R coding language available in print and online. But these tend to be general and abstract, sometimes going on tangents that are not so relevant to what you want to do as a biologist. What makes this book different, is that it is written with the biologist in mind. Specifically, my goal was to write the book that I wish I had had as an undergraduate student learning how to collect and analyze data. With the benefit of hindsight, I've tried to cut out all the programming details that haven't been of much use to me as a data scientist, and focus on the most common methods. I've tried to connect to biological questions and examples as much as possible, without getting too side-tracked with biological details. This decision-making progress is based on my research and teaching experience in a range of topics in Biology and Health Sciences at Queen's University – Environmental Science, Epidemiology, Genomics, Ecology, and Evolution.

A comprehensive coding volume would require thousands of printed pages and take decades to master. In choosing the content for this book, I have focused on everything that I wish I knew when I first started learning to program in R. Many of the functions and packages included here were not available when I started, but have some exceptional functionality. I will continue to add new tricks and techniques that I find useful.

0.1 Is this book for you?

Maybe you are curious about coding for data analysis but you aren't sure if you want to invest the time and energy you will need to become competent in these methods. Many students in biology programs do not receive strong quantitative skills training in math, statistics, or computer science. In fact, many of us choose to go into biology programs because we are scared of the quantitative focus of the 'hard' sciences like physics and chemistry. Only much later do we realize how valuable these skills can be for investigating biological phenomena. Modern biology is defined by 'big data' sources including high-throughput sequencing, real-time environmental measurements, satellite imaging, animal tracking, and monitoring human health. Along with more traditional data types, these data are increasingly made available in online databases that are too big to navigate manually. Coding is not simply helpful to biologists – it's becoming essential.

To help demonstrate the tremendous value of coding, I focus on examples drawn from real biological studies. I try to provide real-world examples of how one can apply programming tools and techniques to curate, analyze, and visualize biological data. These tend to be areas in which I have researched and published papers – opportunities that were presented to me because of my ability to analyze data in a reproducible and open framework. However, a key theme of this book is that these skills are highly transferable, not only across the biological sciences but to other disciplines.

Here are a few examples of the diversity of data, analyses, and visualizations in my own collaborations, which all use similar methods in R:

1. A paper examining rapid evolution of flowering

    https://doi.org/10.1126/science.1242121

2. A *de novo* genome assembly:

    https://doi.org/10.1093/g3journal/jkab339

3. A meta-analysis of evolution of invasive species:

    https://doi.org/10.1111/mec.13162

4. Tracking COVID-19 outbreaks using whole-genome sequencing:

   https://doi.org/10.1038/s41598-021-83355-1

5. A study of metabolites in nasal swabs that can differentiate COVID-19 from other viral infections in human patients:

   https://www.nature.com/articles/s41598-022-14050-y

6. An analysis of 3,429 herbarium images and >1 million weather records to reconstruct evolution of an invasive plant:

   https://www.pnas.org/doi/full/10.1073/pnas.2107584119

7. A model of species range limits:

   https://royalsocietypublishing.org/doi/full/10.1098/rstb.2021.0020

0.2 Why R?

In Biology, there are two dominant programming languages: **Python** and **R**. Thousands of hours have been wasted arguing the merits of one programming language over another. The truth is that there is a lot of similarity and it's very easy to to move from one to the other.

There are many other programming languages used in Biology. **C/C#/C++** and **Java** are popular in computer science because they provide a high level of control, but this comes at a cost of abstraction and a steeper learning curve. Bash programming in **Unix/Linux/GNU** is all but necessary for high-performance computing on remote servers, but in biology it is most often used to automate file management and to run programs written in other languages. **Julia** is gaining momentum for mathematical modellers, but it is still in its infancy. **PERL** was popular for bioinformatics but it has been all but replaced by **Python**.

We focus on **R** because it is more commonly used in published statistical analyses in Biology, and it is a bit easier to learn. As you will see, it is very easy to walk through the fundamentals and generate graphs and statistical analyses with just a few hours of practice. This comes at a cost of slower run-times and less flexibility than Python, but this is usually not a problem for beginners. In fact, it is possible to use R to run Python code (or C++ or many other languages). More importantly, concepts like data objects, function and packages are conceptually very similar between R and Python, making it easy to move from one language to the other. The truth is that they are both good languages and anyone who tells you that language A is better than B is simply showing their ignorance about language B.

0.3 Advice

If you've completed a few years of any undergraduate program in biology, then you've probably developed a good approach to study various subjects in Biology. Maybe it involves reading the textbook, attending lectures, and making notes that you review before the big test. Coding is different.

If this is your first attempt to learn how to code, then it's important to understand HOW to learn to code. You won't learn by reading this textbook. You need to take **participate** and actively take control of your learning by typing along with the examples in this book.

Consider that R is a programming *language*. When I teach this content at the senior undergraduate and junior graduate level, I often begin with a poll of students to see who has learned to speak more than one language. I then ask:

Question: How did you become fluent in a second language?

Some common themes in the answers tend to be:

- Immerse yourself
- Study, read, listen
- Try something new, fail, correct errors, repeat
- Practice, practice, practice!

How do you become fluent in a programming language? Pretty much the same way:

- Immerse yourself
- Study, read, and **type everything out**!
- Try something new, fail, correct errors, repeat
- Practice, practice, practice!

Learning a new language is not easy. Learning a programming language is not easy either. Here are a few specific tips to become fluent in R:

1. **Get organized and PLAN**. Use a personal calender and schedule sufficient time to deal with error messages. This is important to accept, though it can be difficult: troubleshooting your code often takes more time than planning and writing it, especially when you are starting to learn.

2. **Apply what you learn**. You will start to develop a toolbox of coding techniques from day one. Look for opportunities to apply them whenever you can. Try to re-frame small projects or tasks in terms of what you can address with your R toolkit. Even if it takes a lot

longer to code than to use other methods, the extra time will reinforce your coding skills, saving time in the long run. Take time to think about what coding tools you can apply.

3. **Experiment**. Try new things, make mistakes, solve problems.

4. **Devote time**. Set aside large blocks of time (2+ hours), to **immerse yourself** in your coding lessons or project.

5. **Focus**. Eliminate distractions. Turn off your notifications. Put your phone and computer on 'airplane mode'. Do whatever it takes to work without interruption. Get some good headphones with white noise or instrumental sounds (no lyrics) to block out distractions. Here are some things I listen to, depending on mood:

   - Baroque/Classical
   - Smooth Jazz
   - Electronic (ambient, house, lofi)
   - *https://coffitivity.com/*

6. **Learn to Troubleshoot**.

   - If you get stuck, Google: "How do I _____ in R". Look for answers from a website called *Stack Overflow*: *https://stackoverflow.com/*
   - If you can't figure out what an error means, paste it into Google. Again, look for answers from *Stack Overflow*.

7. **Socialize**. Find a coding support group or find a few others to form your own group. Discuss problems and successes. Read other people's code to see how they tackle problems. Rarely is there one single 'right' way to code something.

8. **Git 'er done**. When you are starting out, the 'right' way to code is whatever it takes to get the code to do what you want. Don't let perfection be the enemy of the good: messy code that works is 100% better than efficient code that never runs.

9. **Improve**. As you get more comfortable you can start to think about cleaner, clearer, more efficient ways to code. As you advance, look for ways to do the same thing faster and with fewer lines of code.

10. **Embrace Failure**. I can't stress this enough. Even after 10+ years of programming experience, I often make mistakes, and a decent amount of my time is spent dealing with error messages and unexpected output. Every error is a learning opportunity. This is time well spent.

11. **Read** the documentation for the function or package you are using. Don't worry if you don't understand everything. Be sure to take the time to read it slowly and try to understand as much as you can. Try searching online for terms or phrases that are not familiar to you. You will come across these again in the future, so you are investing time now for future payoff. In addition to the built-in help in R, often the repository on *The Comprehensive R Archinve Network (CRAN)* (*https://cran.r-project.org/*) or *Bioconductor* (*https://www.bioconductor.org/*) will include *vignettes* or *tutorials* as pdf files with worked examples.

0.4 Learn By Doing!

As you work through these self-tutorials, don't just read them. I can't stress this enough: take the time to type out the commands in your R

(Studio) console and make sure you get the same output. The simple act of typing it out will send messages to your brain saying "hey, this is an important thing to remember." If you get an error, even better! Read the error carefully, then compare what you typed to what is in the tutorial. Once you find what is different, you will learn what that error means.

About 70-90% of coding time is dealing with errors, and the same is true for learning to code. This can be difficult for us to accept because our experience in a typical biology course is quite different.

0.5 What to Expect

Learning to code is a lifelong journey. There is always more to learn and new ways to improve. The beginning of your journey might be broken up into three overlapping stages, depending on the level of training you have already received:

1. **Utter bewilderment** – reading code is like reading a foreign language. All these letters and symbols are meaningless to you.

2. **Understanding** – you can look at a function and have a decent idea of what it does and how to use it, but you don't understand most of the parameters. You usually rely on default parameters.

3. **Competence** – you can write your own code from scratch, without needing to look up examples, and you are able to carefully review and apply parameters. You rarely trust default parameters, especially for more complicated functions.

4. **Expertise** – you write your own functions and help others to troubleshoot code, analysis pipelines, etc. Maybe you even have your

own published R package or algorithms.

Don't confuse *understanding* with *competence* – this is a common mistake that students make. It's relatively easy to learn how to understand code that is shown to you, but it's quite another skill to learn the names and parameters of useful functions and apply them to solve problems or answer questions. That doesn't mean you need to memorize every function – though memorization can help. A good strategy to move from understanding to competence is to take the extra time and make the effort to type out the code that is shown to you, even when you can look at it and understand what it does. As noted earlier, the act of typing out the code is what will help to solidify it in your brain.

0.6 Translational Coding

There is often a mismatch between the knowledge acquired through a university degree and the skills that employers need in their workforce. That is, newly minted university students have a lot of knowledge and skills for learning, but often struggle with goals laid out by employers or in entrepreneurial endeavours or thesis/dissertation research.

In the computing world, the disconnect between learning and application can happen when students have acquired knowledge of coding algorithms and tools, but learn to apply these tools only when working within a 'sandbox' created for teaching purposes. The sandbox is a clean and well-groomed programming environment with pre-loaded software and examples, curated by the educator. The sandbox lacks the messiness and ambiguity that define real-world applications, and the student never faces these uncomfortable but highly relevant challenges. The sandbox approach is commonly used in both university settings and online courses (e.g. Udemy, Coursera, Datacamp, Skillshare).

A typical teaching sandbox will probably include pre-installed software with 'clean' data defined by a well-defined data structure without errors or missing observations. It will probably have a clear and singular path from problem to solution. This approach has the advantage of efficiency – both for the educator and for the learner. The learner can be guided to move efficiently through key learning objectives while minimizing unexpected bugs or problems that can slow progress and take significant time for educators to deal with. The sandbox creates a more homogeneous experience that is more efficient for tracking progress and assigning grades. The trade-off is that sandbox learning does a poor job preparing you for the messy realities of coding with real data in the real world.

An alternative to the sandbox approach is *translational coding*, which borrows the term from *translational medicine*. **Translational medicine** is a multidisciplinary hybrid between research and application that directly connects medical researchers to the needs of patients. By analogy, **translational coding** tries to directly connect coding skills and tools to the needs of potential employers.

This will not be pleasant for you, the learner, at first. The sandbox approach is popular with learners because it is relatively quick and painless with minimal time needed for researching, planning, debugging, and other forms of problem solving. There is value to learning to work quickly and efficiently, but there is also value in learning to deal with problems that arise in the real-world. This includes dealing with errors at every stage, from installing software to problems hidden among thousands of lines of data or code. This can be frustrating at first, and it will absolutely slow down your progress. There are three key things to remember when this happens:

1. Every error, problem, or roadblock is a **learning opportunity**. Every problem and assignment will have specific goals and challenges that are explicitly laid out by the tutorial, assignment or practice problem. These are the challenges that every learner must overcome to complete the task. In addition, there are *implicit* challenges that may be unique or shared by only a few learners – a particular typo in the code, an error importing or saving, an unidentified error in your dataset. These implicit problems may feel 'unfair' because not every learner has to deal with the same problems at the same time. Over time however, these will tend to average out so that everyone will make similar mistakes, albeit at different times.

2. You can learn to **budget your time** to deal with these implicit, unforeseen errors. And this is an important and highly-transferrable skill! Start a problem or assignment as soon as possible. Give yourself time to take a break and come back to a problem when you get stuck. When you estimate how long an assignment will take, don't just look at the *explicit* goals. Remember to also add time for the *implicit* challenges, which will take much longer to complete.

3. **Time devoted to a new problem pays off in the future**. Most of your time will be spent the first time you encounter a problem. If you take the time to read the error or warning, think about it, and investigate it, then you will know how to recognize and deal with it in the future. In this way, implicit challenges tend to balance out among learners over time. Some learners will encounter a problem early and struggle while others move ahead, until they encounter the same problem, evening the playing field.

The most important thing is to **embrace the challenge**! Don't let yourself get discouraged.

Now, let's get set up to start coding in R.

Setup

0.7 R

Before you begin these tutorials, you should install the latest version of R: *https://cran.r-project.org/*

Versions are available for Windows, MacOS and Linux operating systems. Immediately we can see one of the advantages of learning to code in R – we can move code across computing platforms quite easily, as long as R is installed there.

0.8 R Studio

You should also install R Studio: *https://rstudio.com/products/rstudio/download/#download*

R Studio is an **Integrated Development Environment (IDE)**. Once you install R Studio, go ahead and run the program.

You will see several helpful *tabs*, probably arranged across four windows. Several windows have more than one tab at the top, which you can click to access. Here is a quick overview of the more useful ones (some of this will make more sense after you work through the first few chapters of the tutorial):

- **Environment** keeps track of all of the objects in your programming environment.
- **History** keeps track of the code you have run.
- **Files** similar to the Finder (MacOS) or File Explorer (Windows), starting with the *working directory*.
- **Plots** are where your plots are created.
- **Packages** show which packages you have installed, and which have been loaded.
- **Help** provides documentation for R functions.
- **Console** is important enough to get its own section.

0.9 Console

The **console** is one of the most important tabs in *R Studio*. It's usually the main tab that opens on the left when you first start R Studio. You'll see a little chevron (>) with a cursor after it. This is the R Console, which is the part of *R Studio* that actually runs the R program. Everything in this window shows you what would happen if you ran the code outside of R Studio, for example on a high perfomance computing cluster like the ones maintained by *Compute Canada*, *Microsoft Azure*, *Amazon Web Services*, or Queen's University's own *Centre for Advanced Computing*. Everything in R studio is built around helping you to perform tasks in R, as shown through the R Console.

0.9 R Script

To run an R script, you can just type functions into the console. However, it is very hard to keep track of everything you do if you only use the console. In R Studio you can click `File-->New File-->R Script`.

This will open a new tab window called **Untitled**. This is called a **script**, but it's really just a text file, with a **.R** suffix, that you can use to keep track of your R program. Try typing something into your R script – don't worry for now if it is just some random text. Note that you can **Save** this file.

Nothing happens (yet). To run the script, you have to send the text from the script tab to the console tab. There are a few ways you could do this:

1. Copy and paste manually. This works fine, but there are more efficient options.

2. Highlight the code you want to run and click the **Run** button on the top-right corner of the script tab. The run button sends the highlighted text from the script to the console.

3. If you click the **Run** button without highlighting text, it will send whatever text is on the same *line* as your cursor.

4. If you press **Ctl + Enter** (Windows) or **Cmd + Return** (Mac) it will do the same thing – this is the shortcut for the **Run** button.

5. There are other options if you press the tiny triangle next to the **Run** button, including **Run All**.

6. **Ctl/Cmd + Shift + Enter/Return** is a shortcut for **Run All**.

0.10 Packages

Packages in R contain functions – small programs that contain functions you can use. A few are loaded automatically when you start R, including the `stats` and base packages. One really good package is called tidyverse. The `tidyverse` package contains a lot of useful functions for working with different types of data, including visualizations. You'll need to make sure you are connected to the internet and that your connection to the internet won't be interrupted during the download.

> WARNING! This may take a long time to run

To install the packages, open R Studio and look for the **Console** tab. Type this into your console:

```
install.packages("tidyverse")
```

Next, install 'devtools:

```
install.packages("devtools")
```

The `install.packages()` function downloads the package and saves it on your computer. You only need to do this one time, though you may want to do it periodically to update to the latest version of the package.

Once a package is installed on your computer, it will be available to run in R with the `library()` command. You'll see examples of this throughout the book.

That's it!

Now, let's get coding...

Chapter 1

R Fundamentals

1.1 Overview

This chapter provides a rapid breakdown of the core functionality of R. There is a lot to cover in a very short time. You may be tempted to skip over some of these sections, but this chapter forms the foundation of future chapters. If you don't have a solid foundation, you will have trouble building your coding skills. Remember that you can only learn coding through repetition. Take the extra time and make the effort to type out each code and run it in your console.

I can't stress this enough: It is important that you physically participate and code along with the examples. Type everything out. The physical act of typing into R and troubleshooting any errors you get is a crucial part of the learning process.

It's very likely you will sometimes get a different result, such as a warning or error message. Don't get frustrated! Think of it as an opportunity to work on you debugging skills. Check to make sure you don't have any typos, like the letter 1 and the number 1, or \ vs /, or missing spaces or other changes that may be hard to spot visually. If you are getting a warning, read it carefully.

1.2 R Basics

Make comments inside your code with the hash mark #. When you type this character, it tells the R program to ignore everything that comes after it.

Documentation is an important part of coding. It takes a bit of extra time to write, but it will save you a lot of time. Careful documentation will be essential when coding collaboratively, even if your collaborator is you when you wrote code six months back.

It's ok to play around with code to get it working, but once you have a piece you are happy with, be sure to go back and add documentation.

Later, we will see how to use R markdown to provide more attractive documents for reproducible analysis. But for dedicated programs, you can get creative with characters to help make long documentation more readable:

```
# Use hastags to make comments - not read by the R console
# Use other characters and blank lines to improve readability:
# -------------------------
# My first R script
# Today's Date
# -------------------------
# Add a summary description of what the script does
# This script will...
# And annotate individual parts of the script
```

1.2 Basic Math

You can do basic mathematical equations in R. Many of us choose to become biologists because we aren't comfortable with mathematical

1.2. R BASICS

equations, only to find out later how important math is for biology! As we'll see later, coding can help to demystify mathematical equations. Let's start with some basics:

Yes, type these out!

```
10 + 2 # add
```

[1] 12

```
10 - 2 # subtract
```

[1] 8

```
10 * 2 # multiply
```

[1] 20

```
10 / 2 # divide
```

[1] 5

```
10 ^ 2 # exponent
```

[1] 100

```
10 %% 2 # modulo
```

[1] 0

Question: Did you type this out? If not, you missed something important. Go back to the beginning of the book and read more carefully.

The modulo %% is one you may not be familiar with, but it comes in really handy in a lot of coding contexts. The modulo is just the remainder of a division. So `10 %% 2` returns a zero because 2 divides into 10 five times, but `10 %% 3` returns a 1 because three divides into 10 three times with 1 remainder.

This can be useful to determine whether a number (x) is even (i.e. if x %% 2 returns zero).

> **Tip**: To get more practice, use R instead of your calculator app whenever you need to calculate something. It seems silly to go through the trouble to open R Studio to calculate a few numbers, but it will get you comfortable using R and R Studio, which will pay off in the long run..

1.2 Objects & Functions

Objects and **functions** are the bread and butter of the R programming language. An object can take many forms, but is generally assigned *from* an *input* or *to* an *output*. This could include a letter or a number, or a set of letters or numbers, or a set of letters and numbers, or more structured types of objects that link together more complex forms of information.

Objects are manipulated with **functions**. Each function has a unique name followed by a set of parentheses, which are used to define input and **parameters** that are used by the function, including inputs and outputs.

In fact, there is a function called `function()`. Yes, there is a function in R called *function*, and you can use it to write your own custom functions, but we'll save that for later.

1.2. R BASICS

For now, just remember that functions have brackets. Brackets are used to define input and parameters that the function uses to produce output.

> **Warning**: Do not put a space between the function name and the opening bracket (or you will generate an error.

1.2 c()

The **concatenate** function c() is a very simple yet important and common function in R. Use it to group items together.

c(1,2,3,5)

[1] 1 2 3 5

In this function, the numbers 1, 2, 3, and 5 are the input parameters. Each number is itself an object in R.

The output is a type of **object** called a **vector** that contains four **elements**. The c() function takes four separate objects (elements) and combines them into a new object (vector). If this seems weird, take a few minutes to think it through because this difference will be important later.

Think of a vector as part of a row or column in a spreadsheet, and an element as one of the cells. We can also have more complex obects that are equivalent to entire spreadsheets, or a combination of multiple spreadsheets and other kinds of structured data.

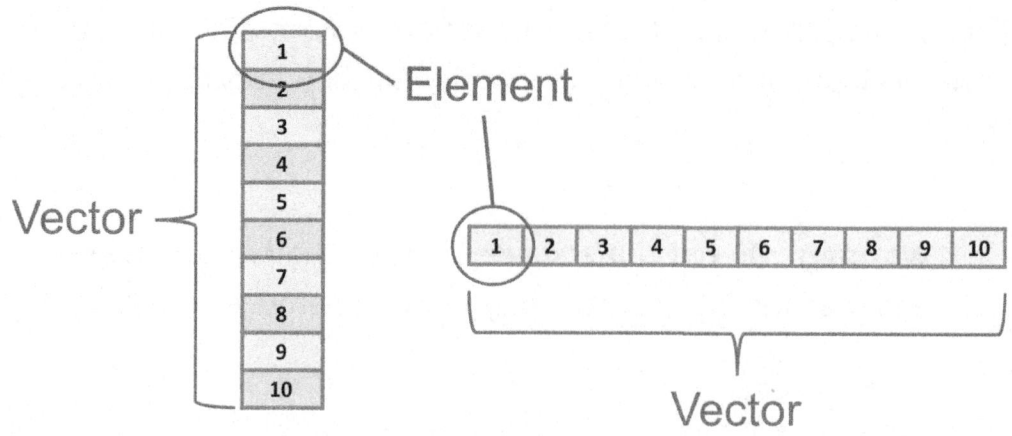

Figure 1.1: Vectors contain elements

1.2 Math Functions

Here are some functions for common mathematical calculations. Type these out and then try channging some of the numbers in brackets to get a feel for them:

```
abs(-10) # absolute value
```

```
[1] 10
```

```
sqrt(10-1) # square root (with subtraction)
```

```
[1] 3
```

```
log(10) # natural log
```

```
[1] 2.302585
```

1.2. R BASICS

```
log10(10) # log base 10
```

[1] 1

```
exp(1) # power of e
```

[1] 2.718282

```
sin(pi/2) # sine function
```

[1] 1

```
asin(1) # inverse sine
```

[1] 1.570796

```
cos(pi) # cosine
```

[1] -1

```
acos(-1) # inverse cosine
```

[1] 3.141593

```
tan(0) # tangent
```

[1] 0

```
atan(0) # inverse tangent
```

[1] 0

Note that `pi` is a special object containing the digits of pi. Try typing `pi` in the R Console and pressing **Enter**.

1.2 Round/Truncate

We can use R for rounding and truncating numbers.

```r
round(pi, digits=3) # standard rounding to 3 digits
```

```
[1] 3.142
```

```r
floor(pi) # round down to closest whole number
```

```
[1] 3
```

```r
ceiling(pi) # round up to closest whole number
```

```
[1] 4
```

```r
signif(pi, digits=2) # round to keep 2 significant digits
```

```
[1] 3.1
```

> **Pro-tip**: `round()` with `digits=3` is a great function to use in your reports, manuscripts, theses, and other scientific documents.

Later, we'll look at how to generate reports that incorporate code (e.g. statistical analyses) that you can quickly update with new data. Rounding the output of your R code with `round()` makes for much cleaner, and more readable reports. More than three digits may be necessary in a few cases, but in most cases it just adds unnecessary clutter.

1.2 Logic Operators

An **operator** is used to compare objects. We'll use these a lot when we start writing our own custom programs and functions. It also comes in handy for sub-setting your data.

```
1 > 2 # greater than
```

[1] FALSE

```
1 < 2 # less than
```

[1] TRUE

```
1 <= 2 # less than or equal to
```

[1] TRUE

```
1 == 1 # equal to
```

[1] TRUE

```
1 == 2 | 1 == 1 # | means 'OR'
```

[1] TRUE

```
1 == 2 & 1 == 1 # & means 'AND'
```

[1] FALSE

```
1 == 1 & 1 == 1
```

```
[1] TRUE
```

We can also use `!` as a negation/inverse operator

```
1 != 1 # not equal to
```

```
[1] FALSE
```

1.2 Group Comparisons

Instead of the vertical bar character `|`, you can use `%in%` with `c()` to check a large number of values.

```
1 %in% c(1,2,3,4,5,6,7,8,9,10)
```

```
[1] TRUE
```

1.2 Congrats!

Before we move on to the next section, take a second to look back at all the coding skills you've already learned: documenting code, basic math, working with objects and functions, combining objects, some advanced math functions, and comparing objects. Well done!

Seriously, you already know enough write your own R program! Try it!

1. Make a new file: `File-->New File-->R Script`
2. Write some code – try to use as many concepts above as you can.
3. Don't forget your documentation!

4. Save the file
5. Run the file and look at the output
6. Debug any errors and warning messages.
7. Show off your program to your friends and family

You are a coder now! Let's take your skills to the next level.

1.3 Use ? for HELP

Whenever you are learning a new function, you should use ? and carefully read about all the parameters and outputs. The explanations can be a bit technical, which is intimidating at first. But after enough practice you will start to understand more and more of the descriptions. Let's break it down:

```
?round
```

> **Note**: In R Studio, the help will open in a separate 'Help' tab (lower, right panel in the default view)

1.3 Description

The description gives a general overview of the function. In this case, `round()` is one of a set of related functions, which are all described together in the same help file

1.3 Usage

This shows the general form of the function that is run in the R Console.

1.3 Arguments

This explains the 'arguments' of the function, which are the input values and parameters. In the case of round the arguments include a numeric vector x as input and `digits` as a parameter.

1.3 Value

This help doesn't have a **Value** subheading, but more complex functions do. For example, try `?lm` to see the help for linear models. Values are objects created by the function as output. For example, the model `coefficients` and `residuals` are separate objects of a linear model created by the `lm()` function.

1.3 Details

This explains the function(s) in greater detail, and is worth reading the first few times you use a function.

1.3 Examples

This section gives examples as reproducible code, which you can copy-paste right into your terminal.

To conclude, always read the help **carefully** when you first use a function. It's normal to keep referring to the help every time you use a function that you aren't too familiar with. It's also normal that you might not understand everything in the help file. Just do your best and be persistent and over time it will start to make more sense to you. You

1.4 Random Numbers

will find these get easier as you read about more functions and try to apply whatever you can understand.

1.4 Random Numbers

The ability to quickly and efficiently generate random numbers has a lot of useful applications for biologists. What are some examples?

1. Generating random numbers as part of an experimental design.
2. Simulating 'noise' or stochastic processes in a biological model.
3. Developing a null model for statistical significance testing.
4. Exploring 'parameter space' in a Maximum Likelihood model or a Markov Chain Monte Carlo simulation.

It is very easy to generate some random numbers in R, from a variety of different sampling distributions.

These are covered in more detail in the *Distributions Chapter* of the book *R STATS Crash Course for Biologists*, which is part of a different book (R Stats Crash Course for Biologists). For now, we'll just focus on generating random numbers.

1.4 Uniform

Perhaps the simplest random number is a whole number (i.e. no decimal) drawn from a **uniform distribution**, meaning that each number has an equal probability of being selected.

```
runif(n=10, min=0, max=1)
```

```
[1] 0.6301161 0.9613978 0.3543704 0.9709359 0.8131712
[6] 0.9699677 0.4500428 0.9810870 0.8844159 0.7548796
```

Note that your randomly chosen numbers will be different from the ones randomly chosen here.

The `runif()` function here uses 3 parameters:

1. `n` – the number of random values to generate
2. `min` – the minimum value that can be chosen
3. `max` – the maximum value that can be chosen.

We'll talk more about parameters later.

1.4 Gaussian

One of the most common random distributions in biology is the **Gaussian distribution** with parameters for `mean` and `sd` (standard deviation). Rational numbers (i.e. with decimal) closer to the mean are more likely to be chosen, with sd defining probability of sampling a value far above or below the mean value.

```
rnorm(10, mean=0, sd=1)
```

```
[1] -0.18032272 -0.84595298 -2.00950111 -0.32663805
[5] -0.57745964  1.20564654 -0.18319917  0.29783508
[9]  0.38959317 -0.09153558
```

Side note: Look what we did here. We wrote 10 instead of n=10 and the function still works! In fact, we can get away with:

```
rnorm(10,0,1)
```

```
[1] -0.79133014 -0.03111041  0.91405471  0.14967977
[5] -0.32826285  1.24381132  1.10836153 -1.24902467
[9] -0.10519823 -0.99250458
```

You can figure out the order by reading the help (?) for the function. When you are starting out, it's a good idea to type the extra characters to specify the parameter names to avoid bugs in your code. It also makes the code more readable to others.

1.4 Poisson

A **poisson distribution** includes only whole numbers with a parameter `lambda`, which is analogous to the mean in the normal distribution.

Poisson distributions are common for count data in biology – seed or egg number, for example.

```
rpois(10, lambda=10)
```

```
[1] 17 12 14  9  7  4  5 11  8  8
```

1.4 Binomial

The **binomial distribution** is useful for binary outcomes – variables with only two possibilities, which can coded as 0 or 1 (or true/false). The `size` parameter is the number of events (e.g. number of coin flips), and the `prob` parameter is the probability of getting a 1 each time.

Binomial distributions are commonly used in population genetics (e.g. sampling alleles with different frequencies).

```
rbinom(10, size=1, prob=0.5)
```

```
[1] 0 1 0 0 1 0 0 1 1 0
```

```
rbinom(10, size=10, prob=0.5)
```

```
[1] 5 8 3 6 4 4 4 6 3 5
```

1.4 Other

Here are a few other random distributions you might be familiar with:

Distribution	R function
Chi-Squared	`chisq()`
t	`t()`
F	`F()`
Exponential	`exp`
Log-Normal	`Lognormal`
Logistic	`Logistic`

1.5 Repeat, Replicate & Sample

In addition to drawing random numbers from defined distributions, it is often helpful to sample from a defined input vector.

For example, maybe we want to generate a data frame with alternating rows for Treatment and Control. We can use the `rep()` function to repeat values.

1.5. REPEAT, REPLICATE & SAMPLE

```
rep(c("Treatment","Control"),3)
```

```
[1] "Treatment" "Control"   "Treatment" "Control"
[5] "Treatment" "Control"
```

Or maybe we want to repeat a function, such as sampling from a normal distribution and calculating the mean of the sample. We could try `rep()` again.

```
rep(mean(rnorm(1000)),3)
```

```
[1] -0.008730267 -0.008730267 -0.008730267
```

Note that your numbers will probably be different, due to random sampling. But there is a problem.

> **Question**: What is wrong with this output?

Answer: We are not repeating the nested function `mean(rnorm())`. Instead, we are just running it once and repeating the output.

To repeat the function, we use `replicate()` instead of `rep()`.

```
replicate(3,mean(rnorm(1000)))
```

```
[1]  0.004131583 -0.009685036 -0.013340741
```

Instead of repeating and replicating, we may want a random sample from our input vector. There are two ways to draw a random sample:

1. **With Replacement** – Randomly sample along the vector and allow for the same element to be sampled more than one. To remember this, imagine each element is a marble in a bag. When your select a specific marble, you *replace* it in the bag so that it can be sampled again.

2. **Without Replacement** – Randomly reshuffle the elements of a vector. Imagine that marbles do not get replaced in the bag, so that each element can be sampled only once.

Sample *with* replacement

```
sample(c(1:10),10,replace=T)
```

[1] 8 6 9 5 9 10 3 1 6 6

Sample *without* replacement

```
sample(c(1:10),10,replace=F)
```

[1] 10 8 3 9 4 6 2 5 7 1

1.6 `set.seed()`

Fun fact: random numbers generated by a computer are not truly random. Instead, the numbers involve a calculation that require a starting number called a **seed**. The seed might be the current Year/Month/Day/Hour/Minute/Second/Millisecond, which means the 'random' number could be determined by somebody who knows the equation and the precise time it was executed.

1.6. SET.SEED()

In practice, computer-generated random numbers are much more 'random' than numbers 'randomly' chosen by a human mind.

We can also take advantage of a computer's pseudo-random number generation by defining the **seed** number. This can help with testing and debugging our code, and for writing code for research that is 100% reproducible. With the same seed, anyone can generate the exact same "random" numbers. We do this with the set.seed() function.

Compare these outputs:

```
runif(5)
```

[1] 0.10281039 0.90757599 0.93853994 0.02910147 0.34884322

```
runif(5)
```

[1] 0.4091996 0.5338017 0.7800766 0.4077009 0.3105005

```
set.seed(3)
runif(5)
```

[1] 0.1680415 0.8075164 0.3849424 0.3277343 0.6021007

```
set.seed(3)
runif(5)
```

[1] 0.1680415 0.8075164 0.3849424 0.3277343 0.6021007

```
set.seed(172834782)
runif(5)
```

[1] 0.13729290 0.18587365 0.01860484 0.88440060 0.21414154

```
set.seed(172834782)
runif(5)
```

```
[1] 0.13729290 0.18587365 0.01860484 0.88440060 0.21414154
```

```
runif(5)
```

```
[1] 0.19787402 0.84870074 0.27303904 0.12225215 0.08365613
```

See how the same 'random' numbers are generated with the same seed?

1.7 Combining objects

Returning now to the concatenation function, we saw how to use use `c()` to concatenate single objects.

```
c(1,2,5)
```

```
[1] 1 2 5
```

We can also *nest* functions, for example we can use `c()` inside of another concatenate function.

```
c(c(1,2,5),c(48,49,50))
```

```
[1]  1  2  5 48 49 50
```

If we need to concatenate a range of whole numbers, we can simplify with the colon :

```
c(1:10)
```

```
[1]  1  2  3  4  5  6  7  8  9 10
```

```
c(100:90)
```

```
[1] 100  99  98  97  96  95  94  93  92  91  90
```

```
c(-1:1)
```

```
[1] -1  0  1
```

Question: How could you use this to generate a set of numbers from -1.0 to 1.0 in increments of 0.1? You already have all the coding knowledge you need to do this! You just have to try combining two of the things you have learned so far.

Hint: Think about how many elements should be in the vector, and what kind of math operation you could use.

1.8 Sequence

Alternatively, you can also use seq() to generate more complicated sequences

```
seq(-1, 1, by = 0.1)
```

```
 [1] -1.0 -0.9 -0.8 -0.7 -0.6 -0.5 -0.4 -0.3 -0.2 -0.1  0.0
[12]  0.1  0.2  0.3  0.4  0.5  0.6  0.7  0.8  0.9  1.0
```

```
seq(-1, 1, length=7)
```

```
[1] -1.0000000 -0.6666667 -0.3333333  0.0000000  0.3333333
[6]  0.6666667  1.0000000
```

1.9 Rows and Columns

As noted above, the output of c() with two or more elements is a **vector** object that is conceptually similar to a set of rows or columns in a spreadsheet.

Use cbind() to bind columns and rbind() to bind rows. The result is a two-dimensional **matrix**, which is conceptually similar to a spreadsheet of n rows by c columns.

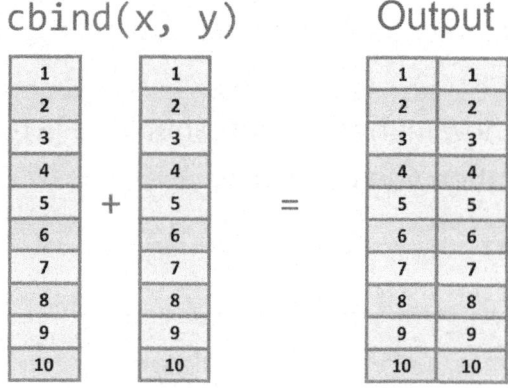

Figure 1.2: cbind() function combines *columns*

```
cbind(1:10,10:1)
```

```
     [,1] [,2]
[1,]    1   10
[2,]    2    9
```

1.9. ROWS AND COLUMNS

```
[3,]    3    8
[4,]    4    7
[5,]    5    6
[6,]    6    5
[7,]    7    4
[8,]    8    3
[9,]    9    2
[10,]  10    1
```

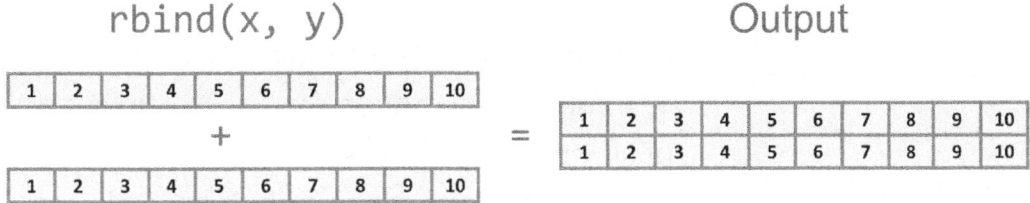

Figure 1.3: cbind() function combines *columns*

```
rbind(1:10,10:1)

     [,1] [,2] [,3] [,4] [,5] [,6] [,7] [,8] [,9] [,10]
[1,]    1    2    3    4    5    6    7    8    9    10
[2,]   10    9    8    7    6    5    4    3    2     1
```

What are n (number of rows) and c (number of columns) for each of the above examples?

1.9 Congrats, again!

Okay, let's take a quick breather from writing code. You have been typing along, right? If not, go back and type out the code. It really is so important if you want to learn this!

We are about to delve deeper into the realm of **object-oriented programming**, but first we need to cover a few basic concepts.

1.10 Data Types

Programming languages like R use different data types.

It's very important to understand data types in order to properly encode and analyze data in R. Here is an overview of the main data types:

Type	Example	Description
string	"String"	Strings are the most common and versatile data type. They can be defined with single ' ' or double " " quotation marks. The downside of strings is that you typically can't do mathematical functions with them.
numeric (**float**)	12.421	Numeric variables are numbers and come in a few flavours. Floats are rational numbers.
numeric (**integer**)	12	Integers are numeric objects that store whole numbers, and may be positive or negative (no decimal).
complex	0+12.43i	Complex numbers include real and imaginary numbers.
logical	T or TRUE	Logical (aka **Boolean**) variables are either TRUE or FALSE, which can be shortened to T and F (**Note** the use of capital letters only). NOTE: TRUE and T are a special *logical* data type and are interchangeable in R, but "TRUE" and 'TRUE' and "T" with quotation marks are *strings* and are treated as separate entities.

1.10. DATA TYPES

Type	Example	Description
factors	any	Factors are a special type of data that may include strings and/or numbers but have a limited number of classes. Factors are often used to code groups in statistical models.

Note that computers cannot store irrational (i.e. infinite, non-repeating) numbers, instead they have encoded as fractions or equations and rounded to some (tiny) decimal place.

Why does it matter? It's very common to have errors in statistical analyses caused by the wrong kind of data. Here is a very common example of a big coding error in Biology:

Imagine you have an experiment set up with three experimental groups coded as 1, 2 and 3.

> **Question**: What data type should these be?

Answer: These should be coded and analyzed as **factors** NOT **numeric** variables. Running statistical anlayses in R on numeric objects that should be factors will give completely different (and wrong!) statistical results.

More generally, you should keep these data types in mind. Consider memorizing them, or even just printing or writing them out and pasting them on your wall. When you get to a point where you are collecting your own data or working with other data sources, you will need to think carefully about which data type each observation should be coded as. This is called **data coding** and it is one of the most important steps in any data analysis pipeline.

1.11 Objects & Variables

R supports **Object-Oriented Programming (OOP)**, which is a programming style that defines and manipulates **objects**

As we have seen, an **object** in R can be a lot of things, but to understand some of the key objects, let's start by thinking about a spreadsheet (example Microsoft Excel).

A spreadsheet has individual cells or elements (boxes) organized into rows (e.g., numbers) and columns (e.g., letters), and may have multiple sheets (tabs). Any of these can be coded objects in R. Objects can also be more complicated types of text files. In biology, we might have DNA (or RNA or protein) sequence data, or matrices of species community data, or time series, or the output of a statistical test. All of these can be coded as objects in R.

Variables are objects that can change value. In R, we can assign variables using <- or =. Almost everything you need to know about R to be a prolific data scientist in biology involves manipulating object *variables* with *functions*!

1.11 Cells (elements)

The most basic object is a single value. For example, a string:

```
X<-"string"
```

Question: Why no output?

Answer: When we wrote: X<-"string" R created the object called **X**. The value of "string" is stored in the R object called X, so no output is produced.

1.11. OBJECTS & VARIABLES

There are a few options to see the contents of X:

```
print(X)
```

```
[1] "string"
```

`print()` Is most generic and versatile for providing feedback while running complex scripts (e.g. during loops, Bash scripts, etc)

```
paste(X)
```

```
[1] "string"
```

`paste()` Converts objects to a string, we'll come back to this.

```
X
```

```
[1] "string"
```

Generally `print()` or `paste()` are preferred over calling the object directly.

OR, we can put the whole thing in brackets, which saves a line of code:

```
(X<-"string")
```

```
[1] "string"
```

Which one should you use? It's ok to use the bracket methods for simple scripts and reports, but use `print()` for more complicated analysis pipelines, especially those that run through a scheduler on remote computers.

1.11 Vector

A vector is a one-dimensional array of cells. This could be part of a row or column in our spreadsheet example.

Each cell within the vector has an 'address' – a number corresponding to the cell ranging from 1 to N, where N is the number of cells.

The number of cells in a vector is called the **length** of the vector.

All items in a vector must be of the same data type. If you mix data types, then the whole vector will be formatted to the most inclusive type. For example, if you include a string with any other format, then the whole vector will be treated as a string:

```
Xvec<-c(1.1829378, X, 1:10, "E", "Computational Biology", 100:90)
Xvec
```

```
 [1] "1.1829378"              "string"
 [3] "1"                      "2"
 [5] "3"                      "4"
 [7] "5"                      "6"
 [9] "7"                      "8"
[11] "9"                      "10"
[13] "E"                      "Computational Biology"
[15] "100"                    "99"
[17] "98"                     "97"
[19] "96"                     "95"
[21] "94"                     "93"
[23] "92"                     "91"
[25] "90"
```

A string is more inclusive because whole numbers can be stored as strings, but there is universally accepted way to store a character string as a whole number.

1.11. OBJECTS & VARIABLES

Similarly, a vector containing integer and rational numbers is a vector of only rational numbers:

```
c(1,2,3,1.23)
```

```
[1] 1.00 2.00 3.00 1.23
```

> **Protip**: A common problem when importing data to R occurs when a column of numeric data includes at least one text value (e.g. "missing" or "< 1"). R will treat the entire column as text rather than numeric values. Watch for this when working with real data!

If you want to mix data types without converting them, you can use a **list** object, which is described later. But first we will need to get comfortable working with the more basic data types.

1.11.2.1 Subset a Vector

Each cell within a vector has a specific address. Just as text message with the correct email address can find its way to your computer, you can find an element in a vector using its address. Remember that in R, addresses start with the number 1 and increase up to the total number of elements.

R uses square brackets [ ] to subset a vector based on the element addresses.

```
Xvec[1]
```

```
[1] "1.1829378"
```

```
Xvec[13]
```

```
[1] "E"
```

```
Xvec[1:3]
```

```
[1] "1.1829378" "string"    "1"
```

1.11 Matrices

A matrix is a 2-D array of cells, equivalent to one sheet in a spreadsheet program. The `matrix()` function can convert a vector to a matrix.

```
Xmat<-matrix(Xvec,nrow=5)
Xmat
```

```
     [,1]          [,2] [,3]                          [,4] [,5]
[1,] "1.1829378"   "4"  "9"                           "99" "94"
[2,] "string"      "5"  "10"                          "98" "93"
[3,] "1"           "6"  "E"                           "97" "92"
[4,] "2"           "7"  "Computational Biology"       "96" "91"
[5,] "3"           "8"  "100"                         "95" "90"
```

Be sure to understand what happened here. Compare this `Xmat` object to the `Xvec` object, above. See how we have re-arranged the elements of a one-dimensional vector into a two-dimensional matrix? **Note**: these two objects need the same number of elements – 1×25 for `Xvec` and 5×5 for `Xmat`.

1.11. OBJECTS & VARIABLES

1.11.3.1 Subset a Matrix

Did you notice the **square brackets** along the top and left side? Do you see how the **rows** have numbers *before* the comma and **columns** have numbers *after* the comma?

These show the address of each element in the matrix. We can subset with square brackets, just like we did with vectors. Since there are two dimensions, we need to specify two numbers using the syntax [row,column].

For example, if we want to select the element from the 3rd column of the 1st row:

Xmat[1,3]

[1] "9"

Or leave it blank if you want the whole row or column:

Xmat[1,]

```
[1] "1.1829378" "4"         "9"         "99"
[5] "94"
```

Xmat[,3]

```
[1] "9"                      "10"
[3] "E"                      "Computational Biology"
[5] "100"
```

> **Protip**: Always remember [row,col]: "**rows** *before* and **columns** *after* the comma. Say it with me "Rows before columns". Say it again, and again, until it is hard-coded in your brain.

1.11 Arrays

Array is a general term to describe any object with N dimensions. We've already seen a few different examples:

Dimension	Object Name
0	Cell
1	Vector
2	Matrix
3+	Array

In R you can build arrays by adding as many dimensions as you need using the `array()` function.

```
Xarray<-array(0, dim=c(3,3,2)) # 3 dimensions
Xarray
```

```
, , 1

     [,1] [,2] [,3]
[1,]   0    0    0
[2,]   0    0    0
[3,]   0    0    0

, , 2

     [,1] [,2] [,3]
[1,]   0    0    0
[2,]   0    0    0
[3,]   0    0    0
```

Notice how 3rd dimension is sliced to print out in 2D. Another way to conceptualize this array is to think of two matrices with the same di-

1.11. OBJECTS & VARIABLES

mension (rows-by-columns). The element of each matrix can be addressed by its [row,col] but we need a third dimension do distinguish between the two matrices. You can see this in the output above each matrix: ,,1 vs ,,2. Together, this array has three dimensions: [row,col,matrix].

> **Question**: If we add a third matrix with the same number of rows and columns, how many dimensions would you need to pull out a specific cell element in R? What if there were 10 or 100 instead of three?

Answer: Three! In each case, we whould still have a 3-dimensional array, and we can access any element as above: [row,col,matrix]. All we are changing is the matrix number from 2 to 3 to 10 to 100!

Higher-order arrays are also possible, but a bit tricky to read on a 2-dimensional screen, and very hard to conceptualize.

Here's an example of a six-dimensional array.

```
Xarray<-array(rnorm(64), dim=c(2,2,2,2,2,2))
```

Once you get the hang of it, it's easy to subset. Just think of each dimension, separated by commas.

```
Xarray[1:2,1:2,1,1,1,1]
```

```
         [,1]       [,2]
[1,] -1.718987  0.3487603
[2,]  1.779268 -0.3523615
```

```
Xarray[1:2,1,1,1:2,1,1]
```

```
          [,1]      [,2]
[1,]  -1.718987 0.8664164
[2,]   1.779268 1.2394975
```

Question: Why are these numbers not the same?

Answer: Look at the `array[]` function and compare to the 6-D array to understand how this works. Each function captures a different 2-dimensional subspace of the 6-dimensional array

If these higher-dimension arrays are too abstract, don't worry. You can get a better understanding with practice. They are important for neural networks, machine learning, and multivariate data (e.g. quantitative genetics, community ecology). Luckily, most of what you need to know you can extrapolate from your intuition about 2-dimensional and 3-dimensional space. Just make sure you understand the similarities and differences among cells/elements, vectors and matrices before you move on.

1.12 Matrix Algebra

R is pretty handy for matrix calculations that would be very time-consuming to do by hand or even in a spreadsheet program.

As an example, let's create some numeric vectors that we can play with. First, a simple vector object called X containing the numbers 1 through 10.

1.12. MATRIX ALGEBRA

```
X<-c(1:10)
X
```

```
[1]  1  2  3  4  5  6  7  8  9 10
```

Second, a vector called Y containing the numbers 0.5 to 5 in 0.5 increments. Note how we can do this using some simple math:

```
Y<-c((1:10)*0.5)
Y
```

```
[1] 0.5 1.0 1.5 2.0 2.5 3.0 3.5 4.0 4.5 5.0
```

Note the extra brackets (1:10) to help us understand that each number in the vector 1 through 10 is multiplied by 0.5, not just the last number.

1.12 Basic Operations

Probably the most common calculation for these X and Y objects is just to cycle through each element of each vector and multiply them together. For example, if X is a vector of leaf length measurements and Y is a vector of leaf width measurements, then we might want to estimate leaf area by multiplying each length by its corresponding width.

In R we just use the standard multiplication operator * on a vector, just like we would do for two individual numbers.

```
X * Y
```

```
[1]  0.5  2.0  4.5  8.0 12.5 18.0 24.5 32.0 40.5 50.0
```

Addition, subtraction, division, and exponents are similar.

```
X + Y
```

```
[1]  1.5  3.0  4.5  6.0  7.5  9.0 10.5 12.0 13.5 15.0
```

```
X / Y
```

```
[1] 2 2 2 2 2 2 2 2 2
```

```
X ^ Y
```

```
[1] 1.000000e+00 2.000000e+00 5.196152e+00 1.600000e+01
[5] 5.590170e+01 2.160000e+02 9.074927e+02 4.096000e+03
[9] 1.968300e+04 1.000000e+05
```

Just as we apply operators to vectors, we can also apply functions to vectors. When we do this, the same function is applied to each individual cell of the vector.

```
log(X)
```

```
[1] 0.0000000 0.6931472 1.0986123 1.3862944 1.6094379
[6] 1.7917595 1.9459101 2.0794415 2.1972246 2.3025851
```

```
exp(Y)
```

```
[1]   1.648721   2.718282   4.481689   7.389056  12.182494
[6]  20.085537  33.115452  54.598150  90.017131 148.413159
```

1.12 Matrix Multiplication

Vectors, matrices, and higher-order arrays have multiple elements. Because of this, there are more than one ways to multiply the elements in one object with the elements in the other. This may seem a bit abstract but matrix multiplication has broad applications in biology, from gene expression and molecular biology to community ecology and image analysis.

There are more options than simply multiplying each corresponding element. For example, we can multiply each element in the vector X by each element in the vector Y. This will create a matrix. Let's make an example with the first 4 elements of X and the first 3 elements of Y.

1.12 Outer product.

In the outer product we work across columns of the first object, multiplying by rows of the second object. It's easier to understand by example:

```
Z<-X[1:4] %o% Y[1:3]
Z
```

```
     [,1] [,2] [,3]
[1,]  0.5   1  1.5
[2,]  1.0   2  3.0
[3,]  1.5   3  4.5
[4,]  2.0   4  6.0
```

Note how the first column is each value of X (1-4) multiplied by the first value of Y (0.5), and the second column is multiplied by the second value of Y (1). Similarly, the first row is each value of Y multiplied by the first value of X (1), etc. What happens if we reverse the order?

```
YoX<-Y[1:3] %o% X[1:4]
YoX
```

```
     [,1] [,2] [,3] [,4]
[1,]  0.5   1   1.5    2
[2,]  1.0   2   3.0    4
[3,]  1.5   3   4.5    6
```

Question: How does the Z matrix object differ from YoX?

Answer: We have switched the rows and columns, which is called a **transpose**

1.12 Transpose

In R, we can transpose matrices with the t() function

```
t(YoX)
```

```
     [,1] [,2] [,3]
[1,]  0.5   1   1.5
[2,]  1.0   2   3.0
[3,]  1.5   3   4.5
[4,]  2.0   4   6.0
```

To multiply two vectors together with the outer product, we arrange the first vector as rows, and the second vector as columns, and then multiply each pair of values together to fill in the matrix.

We can extend this to multiply two objects that are 2-dimensional matrices instead of 1-dimensional vectors. However, this gets tricky for the outer product because instead of generating a 2-D matrix from two 1-D vectors, we will generate a 4-D array from the outer product of two 2-D matrices.

1.12 Dot Product

Another way to multiply two vectors is with the dot product. To do this, we match the element of each **row** in the **first** object with each **column** in the *second* object, and sum them together: (e.g. X[1]*Y[1]+X[2]*Y[2]...).

It's easy to extend from two vectors to two matrices, just by multiplying out elements in each row of the first object by elements in the second object.

```
X %*% Y
```

```
        [,1]
[1,] 192.5
```

```
sum(X*Y) == X %*% Y
```

```
        [,1]
[1,] TRUE
```

1.12 Other Operations

There are a few other important matrix operations that are useful for biological data and modelling/simulations. The **cross-product** is a complicated formula that is easy to calculate in R

```
crossprod(X[1:4],Z) # Cross product
```

```
     [,1] [,2] [,3]
[1,]   15   30   45
```

```
crossprod(Z) # Cross product of Z and t(Z)
```

```
     [,1] [,2] [,3]
[1,]  7.5   15 22.5
[2,] 15.0   30 45.0
[3,] 22.5   45 67.5
```

The **Identity Matrix** is a special matrix with 1 on the diagonal and 0 on the off-diagonal. We can create it with the `diag()` function

```
diag(4) # Identity matrix, 4x4 in this case
```

```
     [,1] [,2] [,3] [,4]
[1,]    1    0    0    0
[2,]    0    1    0    0
[3,]    0    0    1    0
[4,]    0    0    0    1
```

We can also use the `diag()` function on an existing matrix, to pull out all of the values on the diagonal, resulting in a vector

```
diag(Z) # Diagonal elements of Z
```

```
[1] 0.5 2.0 4.5
```

Some of these calculations can get a bit tricky – especially when we move to 2D matrices instead of vectors. You'll want to consult or review a matrix algebra textbook if you are going to apply these, but that's getting too advanced for this book. For now, the important thing is just to know that these options are available if you need them in the future.

1.12.6.1 Matrix Math Summary

Operator	Name
*	Multiply elements
%*%	Dot Product
%o%	Outer product
t()	Transpose
crossprod()	Cross-product
diag(4)	Identity of 4x4 matrix
diag(M)	Diagonal elements of matrix M

1.13 PCA

One popular use-case for matrix calculation is the principal components analysis (PCA). The PCA is covered in much more detail in the *PCA Chapter* in the book *R STATS Crash Course for Biologists*.

Briefly, PCA is a form of **unsupervised machine learning**. It uses matrix math to re-scale a bunch of **correlated** vectors (e.g. measurements) so that they can be mapped to an equal number of **independent** PC axes. For example, if you measure tail fin lengths and body lengths of 100 fish, then you can code the data as two vectors. These values will probably be correlated with bigger fish having bigger tails. We can re-scale these two *dependent* (i.e. correlated) vectors as two *independent* (i.e. uncorrelated) principal components.

In the example shown in the figure, PC1 is a measure of fish AND tail size, whereas PC2 is a measure of tail fin size *relative to* body size.

PCA and similar ordination methods are widely used in biology, from

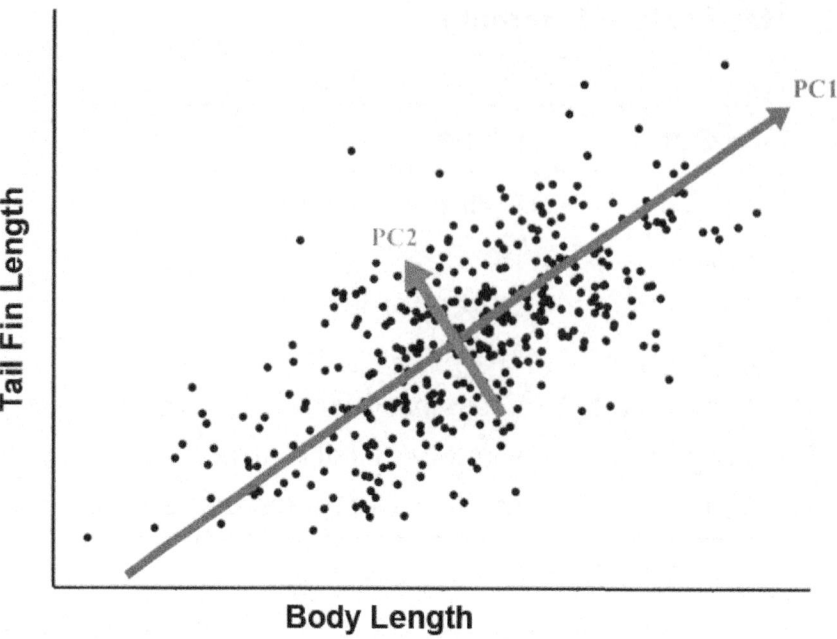

Figure 1.4: PCA of fish size

community ecology and microbiome studies to morphometrics, population genetics, metagenomics and gene expression. Of course there are many applications outside of biology too! For now, just know that it is easy to run a PCA using the `princomp()` function. In most cases, we would want to scale the vectors to have a mean of 0 and standard deviation of 1. Equivalently, we can use the `cor=T` parameter to use the correlation matrix in the calculations.

```
princomp(Z, cor=T)
```

```
Call:
princomp(x = Z, cor = T)

Standard deviations:
    Comp.1       Comp.2       Comp.3
```

```
1.732051e+00 4.214685e-08 0.000000e+00
```

```
3   variables and   4 observations.
```

1.14 Lists

Matrices and higher-order arrays generally all have the same data type and sub-dimension. For example, if you want to combine two separate 2D matrices into a single 3-D array, then the individual matrices have to have the same number of rows and columns. They should also have the same data type, or else everything will be converted to the most inclusive type, as noted earlier in the *Vectors* section.

Often we may want to link different types of information together while still maintaining their different data types. Think of a record in a database where you may have information about an organism's taxonomic classification (factors) height (numeric), weight (numeric), general notes and observations (string), number of scales (integer), and maybe a photograph (numeric matrix) and a DNA sequence (string vector). This wouldn't fit neatly into an array format. Instead, we can use a **list** object.

Lists are useful for mixing data types, and can combine different dimensions of cells, vectors, and higher-order arrays.

Each element in a list needs a name:

```
MyList<-list(name="SWC",potpourri=Xvec,numbers=1:10)
MyList
```

```
$name
[1] "SWC"
```

```
$potpourri
 [1] "1.1829378"           "string"
 [3] "1"                   "2"
 [5] "3"                   "4"
 [7] "5"                   "6"
 [9] "7"                   "8"
[11] "9"                   "10"
[13] "E"                   "Computational Biology"
[15] "100"                 "99"
[17] "98"                  "97"
[19] "96"                  "95"
[21] "94"                  "93"
[23] "92"                  "91"
[25] "90"

$numbers
 [1]  1  2  3  4  5  6  7  8  9 10
```

Important: Many of the statistical functions and other tools in R use list objects to store output. Taking some time now to think about how lists work will save time later when you need to interpret output of R functions.

1.14 Subset

There are a few different ways to subset a list object. We can subset by name using the $ character

```
MyList$numbers # Use $ to subset by name
```

```
[1]  1  2  3  4  5  6  7  8  9 10
```

Or we can **slice** using square brackets.

1.14. LISTS

```
MyList[3] # A 'slice' of MyList
```

```
$numbers
 [1]  1  2  3  4  5  6  7  8  9 10
```

This is similar to the way we used [ ] in vectors and matrices BUT note the inclusion of the name $numbers at the top of the output.

With lists, we have another option, to **extract** using double square brackets.

```
MyList[[3]] # An 'extract' of MyList
```

```
 [1]  1  2  3  4  5  6  7  8  9 10
```

> What's the difference between [ ] and [[ ]]?

This is a bit tricky, but if you invest some time now to understand, you will save yourself a lot of headaches troubleshooting error messages in your code. Do your future-self a favour and take some time to understand this...

First, Look carefully at the output above; notice how the [ ] includes $numbers but the [[ ]] includes only the values? This is important if you want to use the slice:

```
2*MyList[3]
```

```
Error in 2 * MyList[3]: non-numeric argument to binary operator
```

```
2*MyList[[3]]
```

```
[1]  2  4  6  8 10 12 14 16 18 20
```

Note the error generated in the first case

The second case is just a pure vector of numbers, that's why we can multiply each value by two. The first case is still connected to a list object, with the $numbers indicating that we are looking at the *numbers* element of the list. This is part of a larger object, so R returns an error when we try to multiply a number.

In other words, the $numbers heading is part of the sliced object created with [ ] but NOT the extracted object created with [[ ]].

1.14 Function Output

Pro-tip: Many analysis functions in R output as lists, including some very useful functions like lm() for linear models, or princomp() for principal components analysis.

For example, a princomp output list contains several useful objects.

```
princomp(Z)

Call:
princomp(x = Z)

Standard deviations:
      Comp.1         Comp.2         Comp.3
2.091650e+00 2.980232e-08 0.000000e+00

  3  variables and  4  observations.
```

```
names(princomp(Z))
```

```
[1] "sdev"     "loadings" "center"   "scale"    "n.obs"
[6] "scores"   "call"
```

```
princomp(Z)$center
```

```
[1] 1.25 2.50 3.75
```

```
princomp(Z)$scale
```

```
[1] 1 1 1
```

Look at the help ?princomp and scroll down to the **Value** subheading. Note how the subheadings correspond to names(princomp(z))? These values are stored as a list object with each element corresponding to a part of the list object denoted by $.

1.15 print() and paste()

As noted earlier, the print function is the go-to function for printing output to the user. The paste function is useful for combining things together.

Paste is a versatile function for manipulating output:

```
paste("Hello World!") # Basic string
```

```
[1] "Hello World!"
```

```r
paste("Hello","World!") # Concatenate two strings
```

```
[1] "Hello World!"
```

Sometimes we need to convert numbers to strings. `paste` is an easy way to do this:

```r
paste(1:10) # Paste numbers as strings
```

```
 [1] "1"  "2"  "3"  "4"  "5"  "6"  "7"  "8"  "9"  "10"
```

```r
paste(1:10)[4]
```

```
[1] "4"
```

Note how each number above is a separate cell in a vector of strings. Use `as.numeric` to convert strings back to numbers.

```r
as.numeric(paste(1:10)) # Convert back to numbers
```

```
 [1]  1  2  3  4  5  6  7  8  9 10
```

We can also use the `collapse` parameter to condense a multi-cell vector into a single cell. Reading the help for new functions reveals a lot of great treasures like this!

```r
paste(1:10,collapse=".")
```

```
[1] "1.2.3.4.5.6.7.8.9.10"
```

> **Question**: What happens if we paste objects of different length?

```
paste(c("Hello","World"),1:10,sep="-")
```

```
[1] "Hello-1" "World-2" "Hello-3" "World-4" "Hello-5"
[6] "World-6" "Hello-7" "World-8" "Hello-9" "World-10"
```

Answer: The shorter vector gets recycled until every element of the longer vector is pasted.

It is not uncommon to nest a paste function within a print function to communicate output in more complex R scripts.

```
print(paste("Hello",1:10,sep="-"))
```

```
[1] "Hello-1" "Hello-2" "Hello-3" "Hello-4" "Hello-5"
[6] "Hello-6" "Hello-7" "Hello-8" "Hello-9" "Hello-10"
```

This would be useful if you were running a long program with many steps, maybe on a remote high-performance computer where you submit your jobs to a scheduler and you want your program to keep notes about its progress, or generate other notes or feedback. The output of paste is not shown on the screen if used inside of a loop, whereas the output of print is. More about loops below.

1.16 External Files

So far we've done everything within the R environment. If we quit R, then everything we have made will be removed from memory and we'll have to start all over.

For larger projects and reproducible analysis, it is useful to save and load data from external files.

1.16 Working Directory

The **working directory** is the place on your computer where R looks to load or save your files. You can find your current working directory with the `getwd()` function.

```
getwd()
```

Note: The output is specific to your computer, so it isn't shown here.

1.16 Absolute Path

The directory shown in `getwd()` is called an **absolute path**. A path is just computer jargon for the way you get to your working directory, like walking down a path in your computer, turning into the correct directory or folder each time until you reach your destination. The *absolute* term means that it is a location that is unchanging. The problem, for reproducible research, is that the location is specific to your user profile on your computer.

You can set an absolute path with `setwd()`. Here's one example:

```
setwd("C:/Users/ColauttiLab/Documents")
```

> Did you type out the above line? You should if you have been following the instructions! If you haven't, go back to where you stopped and type everything out to reinforce it in your brain. Remember, going through and typing everything out is one of the most effective ways to learn to code.

1.16. EXTERNAL FILES

If you have been typing along, you should have an error message, unless you are working in Windows and for some reason have a *ColauttiLab* username. Now try changing to a different directory on your computer.

If you are a mac user, your directory is probably similar, but without the `C::`

```
setwd("/Users/ColauttiLab/Documents")
```

Again, you will get an error unless you replace the above with a directory that exists on your own computer.

This is why you should **never use absolute path names in your code**. You should always aim for reproducible code, and absolute paths are not reproducible on other computers.

Don't worry, there is a better way...

1.16 Relative Path

In the **absolute path** example above, we first go to the **root** directory, which is the most basic level (or the C: drive in the case of Windows). From the root directory we first click on the **Users** folder, then the **ColauttiLab** folder, and finally the **Documents** folder.

In R, we can just provide a path name as text rather than clicking through all the different folders each time. But as we have seen, the problem with absolute path names is that they are often unique to each user.

Instead of an absolute path, we should use a **relative path** in our code. The relative path in R is denoted with a period, usually followed by a forward slash.

For example, if we have a folder called Data inside our Documents folder, and our current working directory is one of the two examples above, we can use a relative path name to set the Data folder as the working directory. Before you type this out, you should make a folder called Data inside of your current working directory, or else you will get an error.

```
setwd("./Data")
```

The single dot (.) means *inside of my current directory* and the /Data means *move into the Data folder*. This is the coding equivalent of double-clicking the data Folder in your Windows File Explorer or Mac OS Finder.

Now try running this code:

```
getwd()
setwd("..")
getwd()
```

Compare the working directories. The double dot (..) means *go to the parent directory* (i.e. directory containing the current working directory).

The neat thing about relative directories is that it makes it easy to share complex R code between Windows, MacOS and Linux/Unix. In fact, the syntax used by R is the same as Unix, GNU, and Linux.

To make relative paths reproducible, we just have to make sure that the user has all of the relevant files and directors that we are using in our main working directory. R Studio makes it easy to organize your files and code in a sharable working directory by creating an R Project folder.

1.16 R Projects

Working with relative paths can get a little bit confusing, especially if you start using `setwd()`. A good way to avoid confusion is to make an R project in R Studio

`File-->New Project-->New Directory-->New Project`

Then make a name for your project and choose where you want to save it on your computer.

Now quit R studio and look inside the new directory that you just created on your computer. You will see a file with the name of your project followed by `.Rproj`

If you can't see this file, make sure you can view hidden files. Search for 'show hidden files' in your operating system help if you don't know how to do this.

This file is an R project file, and you can double-click on it to open the project. Now here's the cool part: Start R Studio by double-clicking the `.Rproj` *project file* instead of opening the *R Studio App* directly. This should open R Studio, but you will see that the *project folder* will be your default relative path, which you can check with `getwd()`.

There are several good reasons to use R Projects, which become more obvious as you progress as a coder and start working on collaborative projects. For now, think of your project folder as your self-contained programming pipeline. In principle, you want to be able to send the project folder to somebody else to run on another computer without making any changes to the code. You can do this if you learn how to use relative path names.

1.16 Import Data

Download this data file from the Colautti Lab resources website: *https://colauttilab.github.io/RCrashCourse/FallopiaData.csv*

Save this as a `.csv` file in a directory called `Data` inside of your new project folder:

Now open the file with a text editor and take a look at it.

The `.csv` suffix stands for 'Comma Separated Values'. This is really just a regular old text file with different columns separated by commas and different rows separated by each line of text (i.e. hit 'Enter' to add a new row). You can see this if you try opening the file in a simple text editor (e.g. Notepad for Windows or textEdit for MacOS).

> **Pro-tip**: You can easily create a `.csv` by choosing the *Save As* or *Export* in most spreadsheet programs (e.g. MS Excel), and choosing CSV as the output format.

To import this data into R, we can use the `read.csv()` function and save it as an object.

```
MyData<-read.csv("Data/FallopiaData.csv",header=T)
```

Often we have column names as the first row, so we include the parameter `header=T` to convert the first row to column names.

Data without column names would have data on the first row, so we would want `header=F` or else our first row of data would be treated as column names.

Important: In R, objects created by `read.csv` and other `read.?` functions are special objects called `data.frame` objects.

1.16 `data.frame`

A `data.frame` is a special type of object in R that is similar to a 2D matrix, but with additional indexing information for rows and columns of data. This format is partly why base R is so useful for writing a quick, reproducible data analysis.

There are a number of useful functions for inspecting a `data.frame` object.

The indices used for column names can accessed with the `names()` function

```
names(MyData)
```

```
 [1] "PotNum"       "Scenario"    "Nutrients"
 [4] "Taxon"        "Symphytum"   "Silene"
 [7] "Urtica"       "Geranium"    "Geum"
[10] "All_Natives"  "Fallopia"    "Total"
[13] "Pct_Fallopia"
```

There are also a number of functions to quickly inspect the data.frame:

1. Show the first six rows of data

```
head(MyData)
```

```
  PotNum Scenario Nutrients Taxon Symphytum Silene Urtica
1      1        1       low   low japon      9.81  36.36  16.08
2      2        2       low   low japon      8.64  29.65   5.59
3      3        3       low   low japon      2.65  36.03  17.09
4      4        5       low   low japon      1.44  21.43  12.39
5      5        6       low   low japon      9.15  23.90   5.19
6      6        7       low   low japon      6.31  24.40   7.00
```

```
  Geranium Geum All_Natives Fallopia Total Pct_Fallopia
1     4.68 0.12       67.05     0.01 67.06         0.01
2     5.75 0.55       50.18     0.04 50.22         0.08
3     5.13 0.09       60.99     0.09 61.08         0.15
4     5.37 0.31       40.94     0.77 41.71         1.85
5     0.00 0.17       38.41     3.40 41.81         8.13
6     9.05 0.97       47.73     0.54 48.27         1.12
```

2. Show the last six rows

```
tail(MyData)
```

```
    PotNum    Scenario Nutrients Taxon Symphytum Silene
118    143 fluctuations      high bohem      5.06  12.81
119    144 fluctuations      high bohem     19.93  21.07
120    145 fluctuations      high bohem      4.89  32.93
121    147 fluctuations      high bohem      7.84  31.16
122    148 fluctuations      high bohem      4.15  38.70
123    149 fluctuations      high bohem      1.72  10.41
    Urtica Geranium Geum All_Natives Fallopia Total
118  23.82     3.64 0.16       45.49    21.31 66.80
119   6.08     2.80 0.43       50.31     0.00 50.31
120   6.30     9.64 0.00       53.76     2.36 56.12
121  13.61     6.58 0.03       59.22     3.74 62.96
122  23.59     5.11 1.36       72.91     5.89 78.80
123  23.48     8.51 0.43       44.55    19.70 64.25
    Pct_Fallopia
118        31.90
119         0.00
120         4.21
121         5.94
122         7.47
123        30.66
```

3. Check the dimension – the number of rows and columns

1.16. EXTERNAL FILES

```
dim(MyData)
```

```
[1] 123   13
```

4. Check the number of rows only

```
nrow(MyData)
```

```
[1] 123
```

5. Check the number of columns only

```
ncol(MyData)
```

```
[1] 13
```

6. Interrogate the **structure** of the data

```
str(MyData)
```

```
'data.frame':     123 obs. of  13 variables:
 $ PotNum      : int  1 2 3 5 6 7 8 9 10 11 ...
 $ Scenario    : chr  "low" "low" "low" "low" ...
 $ Nutrients   : chr  "low" "low" "low" "low" ...
 $ Taxon       : chr  "japon" "japon" "japon" "japon" ...
 $ Symphytum   : num  9.81 8.64 2.65 1.44 9.15 ...
 $ Silene      : num  36.4 29.6 36 21.4 23.9 ...
 $ Urtica      : num  16.08 5.59 17.09 12.39 5.19 ...
 $ Geranium    : num  4.68 5.75 5.13 5.37 0 9.05 3.51 9.64 7.3 6.36 ...
 $ Geum        : num  0.12 0.55 0.09 0.31 0.17 0.97 0.4 0.01 0.47 0.33 ...
 $ All_Natives : num  67 50.2 61 40.9 38.4 ...
 $ Fallopia    : num  0.01 0.04 0.09 0.77 3.4 0.54 2.05 0.26 0 0 ...
 $ Total       : num  67.1 50.2 61.1 41.7 41.8 ...
 $ Pct_Fallopia: num  0.01 0.08 0.15 1.85 8.13 1.12 3.7 0.61 0 0 ...
```

We can use this to see column headers, types of data contained in each column, and the first few values in each column.

> **Protip**: `str()` is also useful for inspecting other objects, like the output of functions used for statistics or plotting

Pay careful attention to integer `int` vs numeric `num` vs `factor` columns in the `str()` output. These are the data types assigned to each column. As noted earlier, a common source of error students make when starting to analyze data is using the wrong data *type*.

Here's an example of data types gone rogue: In an analysis of variance (ANOVA), you want a `factor` as a predictor and a `num` or `int` as a response. But in linear regression you want `int` or `num` as a predictor instead of `factor`. If you code your factor (e.g. treatment) as a number (e.g. 1-4) then R will treat it as an integer when you import the data. When you run a linear model with the `lm` function, you will be running a regression rather than ANOVA! As a result, you will estimate a slope rather than the difference between group means.

1.16.6.1 Summary

Always check your data types (e.g. using `str`) when you first import the data.

1.16 Subset

The `data.frame` object can be subset, just like a matrix object.

```r
MyData[1,] # Returns first row of data.frame
```

```
  PotNum Scenario Nutrients Taxon Symphytum Silene Urtica
1      1      low       low japon      9.81  36.36  16.08
  Geranium Geum All_Natives Fallopia Total Pct_Fallopia
1     4.68 0.12       67.05     0.01 67.06         0.01
```

```r
MyData[1,1] # Returns first value of data.frame
```

```
[1] 1
```

In addition to numbers, you can subset a column by its header.

```r
MyData[1:4,"PotNum"] # Returns values in "PotNum" column
```

```
[1] 1 2 3 5
```

```r
MyData$PotNum[1:4] # A shortcut to subset the column
```

```
[1] 1 2 3 5
```

Note how we also include 1:4 to show only the first 4 elements, which reduces the output to a more manageable level. If you aren't sure why, try running the above without 1:4 to see the difference.

We can also subset the data based on particular row values. For example, we can find only the records in the *extreme* treatment scenario.

```r
subset(MyData,Scenario=="low" & Total > 60) # Subset
```

```
  PotNum Scenario Nutrients Taxon Symphytum Silene Urtica
1      1      low       low japon      9.81  36.36  16.08
3      3      low       low japon      2.65  36.03  17.09
  Geranium Geum All_Natives Fallopia Total Pct_Fallopia
1     4.68 0.12       67.05     0.01 67.06         0.01
3     5.13 0.09       60.99     0.09 61.08         0.15
```

1.16 New Columns

It's easy to add new columns to a data frame. For example, to add a new column that is the sum of two others:

```
MyData$NewTotal<-MyData$Symphytum + MyData$Silene + MyData$Urtica
names(MyData)
```

```
 [1] "PotNum"       "Scenario"    "Nutrients"
 [4] "Taxon"        "Symphytum"   "Silene"
 [7] "Urtica"       "Geranium"    "Geum"
[10] "All_Natives"  "Fallopia"    "Total"
[13] "Pct_Fallopia" "NewTotal"
```

Notice the new column added to the end. Let's look at the first 10 values:

```
print(MyData$NewTotal[1:10])
```

```
 [1] 62.25 43.88 55.77 35.26 38.24 37.71 49.46 32.77 45.76
[10] 39.20
```

1.17 Other Functions

Here are a few more useful functions for inspecting your data:

1.17 unique()

Find all the unique values within a vector using unique().

```
unique(MyData$Nutrients)
```

```
[1] "low"  "high"
```

1.17 `duplicated()`

Look at each value in a vector and return a TRUE if it is duplicated and FALSE if it is unique.

```
duplicated(MyData$Nutrients)
```

```
  [1] FALSE  TRUE  TRUE  TRUE  TRUE  TRUE  TRUE  TRUE  TRUE
 [10]  TRUE  TRUE  TRUE  TRUE  TRUE  TRUE  TRUE  TRUE  TRUE
 [19]  TRUE  TRUE  TRUE  TRUE  TRUE  TRUE  TRUE FALSE  TRUE
 [28]  TRUE  TRUE  TRUE  TRUE  TRUE  TRUE  TRUE  TRUE  TRUE
 [37]  TRUE  TRUE  TRUE  TRUE  TRUE  TRUE  TRUE  TRUE  TRUE
 [46]  TRUE  TRUE  TRUE  TRUE  TRUE  TRUE  TRUE  TRUE  TRUE
 [55]  TRUE  TRUE  TRUE  TRUE  TRUE  TRUE  TRUE  TRUE  TRUE
 [64]  TRUE  TRUE  TRUE  TRUE  TRUE  TRUE  TRUE  TRUE  TRUE
 [73]  TRUE  TRUE  TRUE  TRUE  TRUE  TRUE  TRUE  TRUE  TRUE
 [82]  TRUE  TRUE  TRUE  TRUE  TRUE  TRUE  TRUE  TRUE  TRUE
 [91]  TRUE  TRUE  TRUE  TRUE  TRUE  TRUE  TRUE  TRUE  TRUE
[100]  TRUE  TRUE  TRUE  TRUE  TRUE  TRUE  TRUE  TRUE  TRUE
[109]  TRUE  TRUE  TRUE  TRUE  TRUE  TRUE  TRUE  TRUE  TRUE
[118]  TRUE  TRUE  TRUE  TRUE  TRUE  TRUE
```

1.17 `aggregate()`

Quickly calculate means of one column of data (`NewTotal`) for each value of another column with groups (`Nutrients`).

```
aggregate(MyData$NewTotal,list(MyData$Nutrients), mean)
```

```
  Group.1        x
1    high 46.51173
2     low 42.76800
```

The *tilde* (~) provides an alternative way to write this function. In R the tilde usually means *by* and it is often used in statistical models.

```
aggregate(NewTotal ~ Nutrients, data=MyData, mean)
```

```
  Nutrients NewTotal
1      high 46.51173
2       low 42.76800
```

> **Hint**: If you got an error, make sure you entered a tilde, not a minus sign.

For the code above, we can say "aggregate NewTotal *by* Nutrients grouping".

The nice thing about doing it this way is that we preserve the column name. Compare the column names here: *Nutrients* and *NewTotal*, vs above: *Group.1* and *x*.

We can also use the colon (:) or asterisk (*) to calculate means across different combinations of two or more grouping columns.

```
aggregate(NewTotal ~ Nutrients:Taxon:Scenario, data=MyData, mean)
```

```
   Nutrients Taxon    Scenario NewTotal
1       high bohem     extreme 45.24833
2       high japon     extreme 45.31500
3       high bohem fluctuations 43.89545
4       high japon fluctuations 44.77692
5       high bohem     gradual 45.36923
6       high japon     gradual 50.43417
7       high bohem        high 52.04273
8       high japon        high 45.69417
9        low bohem         low 41.75231
10       low japon         low 43.86833
```

1.17. OTHER FUNCTIONS

Note that mean in the `aggregate` function is just the `mean()` function in R, applied inside the aggregate function. Instead of mean, we can use other functions, like the standard deviation `sd`:

```
aggregate(NewTotal ~ Nutrients, data = MyData, sd)
```

```
  Nutrients  NewTotal
1      high 11.175885
2       low  8.227402
```

1.17 `tapply()`

`tapply()` works similarly, but using a `list()` function

For example, we can calculate means of each `Nutrients` group:

```
tapply(MyData$NewTotal, list(MyData$Nutrients), mean)
```

```
    high      low
46.51173 42.76800
```

Compare this output with `aggregate` above. Here, the groups are the column names.

1.17 `sapply()` & `lapply()`

The `sapply()` and `lapply()` functions are similar in principle to `tapply()`, but are used to apply a function repeatedly and output the result as a vector (`sapply`) or list object (`lapply`).

Here's an example, where we can summarize the `class` of each column in our `data.frame`

```
lapply(MyData, class)[1:3]
```

```
$PotNum
[1] "integer"

$Scenario
[1] "character"

$Nutrients
[1] "character"
```

```
sapply(MyData, class)[1:3]
```

```
    PotNum   Scenario   Nutrients
 "integer" "character" "character"
```

Compare the above with:

```
class(MyData)
```

```
[1] "data.frame"
```

```
class(MyData$Taxon)
```

```
[1] "character"
```

1.18 Tidyverse

Most of the methods above for managing and summarizing data are the *classic* or *base R* functions. More recently, the **tidyverse** group of functions has gained popularity and these functions have a lot of advantages over the classic tools, particularly for complex data management.

1.19 Save

Just as we can load FROM external files, we can save TO external files. We just change read to write. For a CSV file:

```
## Calulate means
NutrientMeans<-tapply(MyData$NewTotal,list(MyData$Nutrients),mean)
## Save means as .csv file
write.csv(NutrientMeans,"MyData_Nutrient_Means.csv",
          row.names=F)
```

> **Note**: the default for write.csv() adds a column of row names (i.e. numbers) to the output file. To avoid this, use row.names=F

You should see a file called MyData_Nutrients_Means.csv in your working directory.

1.19 Output Foler

Larger projects may generate a lot of different output files, which you may want to organize in an Output folder inside of your project folder. Saving to this folder is easy for relative path names, just add ./Data/ before the file name in your write.csv() function. Just make sure the folder exists before you try to save to it!

```
write.csv(NutrientMeans,"./Data/MyData_Nutrient_Means.csv", row.names=F)
```

1.20 Packages

As noted earlier, **functions** in R use brackets ( ) and generally have **input** and **output** objects as well as **arguments** (i.e. parameters) that affect the behaviour of the functions.

All of the functions in this tutorial are automatically loaded when you start *R* (or *R Studio*), but there are many more functions available. For example, our lab developed the baRcodeR package for creating unique identifier codes with printable barcodes and data sheets to help with sample management and data collection. You may find this helpful for labelling and tracking samples in your own work: *https://doi.org/10.1111/2041-210X.13405*

A **package** in R is a set of functions grouped together. For example, the `stats` package is automatically loaded when you run R and contains many useful functions. You can see what package a function belongs to at the beginning of the help file:

```
?cor
```

The package is shown in curly brackets at the top of the help file. In this case, we see `cor {stats}` telling us that the `cor` function is part of the `stats` package. You can see which packages are loaded if you click on the **Packages** tab in **R Studio** (by default it is in the bottom-right window). The loaded packages are shown with check marks.

1.20 Installing

There are many more packages available that are not yet installed on your computer. You will need to install a new package before you can use it. You only have to do this once, but it is a good idea to update the package periodically, especially when you update to a new version of R. This ensures that you are using the most recent version of the package.

Packages are installed with `install.packages()`, with the package indicated with single or double quotation marks. When you run this code the first time, you may be prompted to choose a repository, in which case choose one that is geographically close to you.

```
install.packages('baRcodeR')
```

Note that **installing** a package just downloads it from an online **repository** (remote computer) and saves to your personal computer.

1.20 Loading

To use a package that you already have installed, you can access it two ways.

1. You can load the package using the `library()` function, giving you access to all of the functions contained within it:

```
library(baRcodeR)
make_labels()
```

This will generate a pop-up menu for creating barcode labels with unique identifiers.

This method is more common, especially when you will use multiple functions from the same package, or use the same functions multiple times.

2. You can use double colons (::) to call a function without loading the whole package

```
baRcodeR::make_labels()
```

This line of code translates to "Run the `make_labels` function from the baRcodeR library". This method is convenient if you just want to use one function from a large library.

> Don't forget to install a package that you want to use by either method.

Another reason to go with the second method is that some packages have **functions with the same names**. Let's say you load two packages pkgA and pkgB that have different functions but both are called cor. When you run the cor function, R will assume you want the one from whichever package was most recently loaded using the `library()` function.

To avoid confusion, you can use the second method to specify which function to run:

```
pkgA::cor()
pkgB::cor()
```

> Did you run the above code? You should still be typing along, in which case you will see the error created when you try to use a package that is not installed on your computer.

1.20.2.1 Library vs Package

The terms **library** and **package** are often used interchangeably. Technically, the **package** is the collection of functions whereas the **library** is the specific folder where the R packages are stored. A library may contain more than one package.

For the most part, you just need to know that a package and a library are a collection of functions.

1.21 Readable code

Now that you've learned how to code, let's take a few minutes to think about best practices. It's important to make your code readable and interpretable by collaborators, peer reviewers, and yourself 6 months from now. There are lots of opinions on this but here are a few basic suggestions:

1. Add documentation to explain what you are doing
2. Add spacing between parameters to improve readability
3. Add spacing on either side of <- when making objects
4. Break long functions into multiple lines; add the line break after a comma
5. Include a list of variables and an overview at the top of your code, or in a separate file for larger projects.
6. Follow these additional suggestions for names:

   - Try to keep your names short and concise but meaningful
   - Use underscore _ or capital letters in your object names to improve readability

- Always start object names with a letter, never a number or symbol
- Avoid symbols completely

Bad	Good
`sum(X,na.rm=T)`	`sum(X, na.rm=T)`
`X`	`Mass`
`Days.To.First.Flower`	`Flwr_Days` or `FDays`
`10d.Height`	`Ht10d`
`Length*Width`	`LxW`

Break up longer code across multiple lines:

Bad:

```
MyObject<-cor(c(1,2,NA,5,9,8,1,2,5),c(2,2,6,3,6,8,3,NA),...)
```

Good:

```
MyObject <- cor(c(1,2,NA,5,9,8,1,2,5),
                c(2,2,6,3,6,8,3,NA),
                method="spearman", na.rm=T)
```

To take your code to the next level, look into the *Tidyverse Style Guide*: *https://style.tidyverse.org/index.html*

Chapter 2

Flow Control

2.1 Overview

Think of your data analysis as a stream flowing from the raw data at the headwaters down to the river mouth, exiting as a full analysis with graphics, statistical analyses, and biological interpretation.

There are different ways we can control the flow of our code. The simplest is just to write a sequence of lines of code, with the output of one line of code forming the input of the next. A pseudo-code example might be:

```
A<-functionA()
B<-functionB(A)
C<-functionC(B)
```

But sometimes we may want to do the same function or analysis only if the input meets certain criteria. Or we may want to reiterate the same analysis multiple times on different inputs. This is where more advanced flow control comes in handy.

To start, let's make up a couple of objects to play with:

```
X<-21
Xvec<-c(1:10,"string")
```

2.2 `if(){}`

The `if(){}` statement uses an **operator** (see above) to asses whether the value is TRUE or FALSE:

```
if(X > 100){ # Is X greater than 100?
  print("X > 100") # If TRUE
} else {
  print("X <= 100") # If FALSE
}
```

```
[1] "X <= 100"
```

A common 'rookie' mistake is to leave out a bracket or use the wrong type of bracket. Use regular brackets for the *if function* `if()` followed by two sets of curly brackets containing the code to run {run if true}else{run if false}.

Break up across multiple lines to improve readability. Note that you don't need an `else{}` part if you just want to do nothing when FALSE.

```
if(X > 0){print ("yup")}
```

```
[1] "yup"
```

2.3 `ifelse()`

The `ifelse()` is a more compact version for simple comparisons. The following code does the same as above.

```
ifelse(X > 100,"X > 100", "X <= 100")
```

```
[1] "X <= 100"
```

2.4 nested `if`

You can also nest `if` and `ifelse` statements to account for more outcomes. Conceptually think of it as a bifurcating tree, starting at the top (root) and then splitting in two for every `if` statement.

```
if(X > 100){
   print("X > 100")
   if(X > 200){
      print("X > 200")
   }
} else {
   if(X == 100){
      print("X = 100")
   } else {
      print("X < 100")
   }
}
```

```
[1] "X < 100"
```

Don't get intimidated. It just takes time to work through all of the possibilities. Try to draw a bifurcating diagram to represent each true/false outcome for the above code.

2.5 for loop

A loop does the same thing over and over again until some condition is met. In the case of a for loop, we set a 'counter' variable and loop through each value of the counter variable. Here are a few examples:

1. Loop through numbers from 1 to 5

```
for (i in 1:5){
    print(paste(X,i,sep=":"))
}
```

[1] "21:1"
[1] "21:2"
[1] "21:3"
[1] "21:4"
[1] "21:5"

2. Loop through the elements of a vector directly

```
for (i in Xvec){
    print(i)
}
```

[1] "1"
[1] "2"
[1] "3"
[1] "4"
[1] "5"
[1] "6"
[1] "7"
[1] "8"
[1] "9"
[1] "10"
[1] "string"

3. Use an index object to index the elements of a vector

```
for (i in 1:length(Xvec)){
  print(Xvec[i])
}
```

```
[1] "1"
[1] "2"
[1] "3"
[1] "4"
[1] "5"
[1] "6"
[1] "7"
[1] "8"
[1] "9"
[1] "10"
[1] "string"
```

Note that in each case there is a vector and the loop goes through each cell in the vector. The `i` variable is an object that gets replaced with a new number in each iteration of the loop.

Loops can be tricky, and the only way to really learn them is to practice as much as possible. Whenever you find yourself writing similar code more than 2 or 3 times, challenge yourself to try to re-write it as a loop.

In addition to looping through a vector, it can often be useful to include a counter variable. This can be especially useful for more complicated loops, but be careful to decide where in your code to update the counter variable. USUALLY it will be either

1. At the **end**, setting the *initial value to 1* before the loop begins.

```
count1<-1
count10<-1

for(i in 1:5){
   print(paste("count1 =",count1))
   print(paste("count10 =",count10))
   count1<-count1+1
   count10<-count10*10
}
```

```
[1] "count1 = 1"
[1] "count10 = 1"
[1] "count1 = 2"
[1] "count10 = 10"
[1] "count1 = 3"
[1] "count10 = 100"
[1] "count1 = 4"
[1] "count10 = 1000"
[1] "count1 = 5"
[1] "count10 = 10000"
```

2. At the **start**, setting the *initial value to 0* before the loop begins.

```
countbefore<-0
countafter<-0

for(i in 1:5){
   countbefore<-countbefore+1
   print(paste("before =",countbefore))
   print(paste("after =",countafter))
   countafter<-countafter+1
}
```

```
[1] "before = 1"
```

2.6. NESTED LOOPS

```
[1] "after = 0"
[1] "before = 2"
[1] "after = 1"
[1] "before = 3"
[1] "after = 2"
[1] "before = 4"
[1] "after = 3"
[1] "before = 5"
[1] "after = 4"
```

This is yet another example of how two different coding approaches can produce the same result.

Read through the outputs above carefully to make sure you understand how the loops work. When you are confident you understand, then write a new loop and write down the predicted output. Run the loop to check if you were right.

2.6 Nested Loops

Counters are particularly valuable when you have a nested loop, which is just one loop inside of another. But note that this can complicate decisions about where to place your counter variable.

In the example below, we are first looping through a vector of length 3, tracked with `i`. Then **for each i** we do a second loop, tracked by `j`.

This time, try to predict the output BEFORE you run the loop. Write it down, then run the loop to check your answer.

```
LoopCount<-0
for(i in 1:3){
  for(j in 1:2){
    LoopCount<-LoopCount+1
```

```
    print(paste("Loop =",LoopCount))
    print(paste("i = ",i))
    print(paste("j = ",j))
  }
}
```

```
[1] "Loop = 1"
[1] "i =  1"
[1] "j =  1"
[1] "Loop = 2"
[1] "i =  1"
[1] "j =  2"
[1] "Loop = 3"
[1] "i =  2"
[1] "j =  1"
[1] "Loop = 4"
[1] "i =  2"
[1] "j =  2"
[1] "Loop = 5"
[1] "i =  3"
[1] "j =  1"
[1] "Loop = 6"
[1] "i =  3"
[1] "j =  2"
```

2.7 `while` loop

The `while` function is another kind of loop, but instead of looping through a predefined set of variables, we iterate until some condition is met inside of the loop. This is called the **exit condition**.

In biology, the `while` loop is often used in optimization simulations, where many calculations are run until some optimum or threshold

2.7. WHILE LOOP

value is reached. Examples may include equilibrium simulations like the Evolutionarily Stable Strategy (ESS), population growth trajectories, or mutation-selection equilibrium. Other examples may include advanced statistical analyses based on maximizing the likelihood or fit of particular statistical models.

One common coding error associated with while loops is that the exit condition is never reached, causing your computer to run an infinite loop.

Here's a simple while loop, which will continue until count is greater than or equal to X.

```
count<-0
while(count < X){
  print(count)
  count<-count+1
}
```

[1] 0
[1] 1
[1] 2
[1] 3
[1] 4
[1] 5
[1] 6
[1] 7
[1] 8
[1] 9
[1] 10
[1] 11
[1] 12
[1] 13
[1] 14
[1] 15

```
[1] 16
[1] 17
[1] 18
[1] 19
[1] 20
```

2.8 Modulo

Earlier in this chapter, we saw the modulo (%%) function, which returns the remainder of a division equation. This comes in handy for loops. For example, if you want to do something every 2, or 3, or N iterations of a loop, you can divide by N and determine if the dividend is zero. Here's an example:

```
for(i in 1:9){
   if(i %% 3 ==0 ){
      print(paste("Iteration:",i))
   }
}
```

```
[1] "Iteration: 3"
[1] "Iteration: 6"
[1] "Iteration: 9"
```

Question: What will the output look like?

Before you run the code, take time to work through the `for` loop and the `if` statement to predict the output. This will help you develop a better understanding of these tricky but highly useful coding tools.

2.9 Faster loops

There are many cases where you may want to add or change vectors, arrays, or data frame objects inside of a loop. By default, R will create a new object each time you add a new element. This can make loops very slow.

Slow loops are usually fine, especially when you are starting out. Does it really matter if your code takes 0.2 seconds or 0.12 seconds? Even if it takes 5 minutes to run, that might be a better use of your time than spending 2 hours finding a faster version.

However, as you advance to code larger projects, you will find that these time differences start to become important. There are some good tricks in R for faster.

This subsection is going to get advanced pretty quickly. Don't worry if you struggle to understand it. Just give it a shot and if you need to move on, file it away for the future. Then, come back to this part of the book if find yourself struggling with loops that are taking too long to run.

Here is a **slow loop** example demonstrating the central limit theorem. First, we want to sample 1000 numbers from a random normal distribution and calculate the average. Second, we want to repeat this process for 5000 iterations. Finally, we want to calculate the average of these 500 iterations. That is, the mean of means.

```
Iters<-500 # Number of iterations
OutVector<-NA
Start<-Sys.time()
for(i in 1:Iters){
  TempMean<-NA
  for(j in 1:1000){ # One loop per sample
    TempMean[j]<-rnorm(1)
```

```
    }
    OutVector[i]<-mean(TempMean)
}
Sys.time()-Start
```

```
Time difference of 1.535169 secs
```

```
paste("Mean of means =",mean(OutVector))
```

```
[1] "Mean of means = -0.00211072942270112"
```

Note the use of `Sys.time()` to keep track of the loop run time. Also, note that the *Time difference* for your computer will depend on the specifications of your computer and memory use. This is a handy technique for large bioinfomatics projects. For example, imagine looping through millions or billions of DNA sequences. How long will the loop take? Try running it on a subset of a few thousand and then multiply to estimate the total run time.

Here is an example of the same calculation in a **fast loop**. Note the only changes are the addition of the `vector()` functions.

```
Iters<-500 # Number of iterations
OutVector<-vector("numeric",Iters)
Start<-Sys.time()
for(i in 1:Iters){
    TempMean<-vector("numeric",1000)
    for(j in 1:100){ # One loop per sample
        TempMean[j]<-rnorm(1)
    }
    OutVector[i]<-mean(TempMean)
}
Sys.time()-Start
```

2.9. FASTER LOOPS

```
Time difference of 0.1519151 secs
```

```
paste("Mean of means =",mean(OutVector))
```

```
[1] "Mean of means = 8.26881764730292e-05"
```

The reason this is faster is a bit technical, but the key is that we are pre-defining the size of our output vectors before we run the loop. This allows R to assign an appropriate amount of computer memory to keep track of changes to the output vector. If we don't do this, R has to constantly update the output memory by creating new objects each time.

But there is an even faster way to do the same calculation. Often if we are outputting to vectors in a for loop, then there is a way to re-write the same function by using sapply() or tapply().

Here is an example of an even **faster loop**.

```
Iters<-500 # Number of iterations
OutVector<-vector("numeric",Iters)
OutMean<-function(x){
  return(mean(rnorm(1000)))
}
Start<-Sys.time()
OutVector<-sapply(OutVector,FUN=OutMean)
Sys.time()-Start
```

```
Time difference of 0.04392195 secs
```

```
paste(mean(OutVector))
```

```
[1] "0.000264860429522105"
```

We've used three tricks here. First, we simply generate a vector of 1000 and take the mean with `mean(rnorm(1000))`, instead of making a nested loop with a vector to hold each of our 1000 random numbers. Second, we put this into a custom function called `OutMean`, which allows us to apply the same function along a vector. We'll look at custom functions in more detail in a later chapter. Finally, we use the `sapply` function to apply our custom function `OutMean` for each element of `OutVector`.

Chapter 3

Quick Visualizations

3.1 Overview

Visualizing data is a key step in any analysis. R provides powerful and flexible graphing tools, whether you are just starting to investigate the structure of your data, or polishing off the perfect figure for publication in a 'tabloid' journal like *Science* (*https://www.science.org/journal/science*) or *Nature*(*https://www.nature.com/*).

In this tutorial, you will learn how to make quick graphs with the `ggplot()` function from the `ggplot2` package. We will go over some options for customizing the look and layout that will allow you to produce professional-grade graphs. In the next chapter you will learn some additional tricks and resources for developing your graphing skills even further.

By the time you are finished these two self-tutorials, you will have all the resources you need to make **publication-ready graphics**!

Both `ggplot` functions come from the `ggplot2` package. It was developed by R Superstar Hadley Wickham and the team at `posit.co`, who also made *R Studio*, *Quarto*, `shiny`, `dplyr`, and the rest of the *Tidyverse* (*https://www.tidyverse.org/*) universe of helpful R packages.

> **Question:** Shat happened to ggplot1?

Once you have mastered these tutorials, you might want to continue to expand your `ggplot` repertoire by reading through additional examples in the `ggplot2` documentation: *http://docs.ggplot2.org/current/*

WARNING: There is a learning curve for graphing in R! Learning visualizations in R can feel like a struggle at first, and you may ask yourself: *Is it worth my time?*

If you already have experience making figures with point-and-click graphics programs, you may ask yourself: *Why deal with all these coding errors when I can just generate a quick figure in a different program?*

There are a few good reasons to invest the time to get over the learning curve and use R for all your graphing needs.

1. You will get much faster with practice.
2. You have much more control over every aspect of your figure.
3. Your visualization will be **reproducible**, meaning anyone with the data and the code can reproduce every aspect of your figure, right down to each individual data point the formatting of each axis label.

The third point is worth some extra thought. Everybody makes mistakes, whether you are graphing with R or a point-and-click graphics program. If you make a mistake in a point-and-click program, you may produce a graph that is incorrect with no way to check! If you make an error in R you will either get an error message telling you, or you will have reproducible code that you can share with somebody who can

3.2. GRAPHICAL CONCEPTS

check your work – even if that *somebody* is you, six months in the future. In this way, **reproducibility is quality** – an attractive but non-reproducible figure is a low-quality figure.

In addition to quality and scientific rigour, there is a more practical reason to value reproducible code. Consider what happens as you collect new data or find a mistake in your original data that needs to be corrected. With a point-and-click program you have to make the graph all over again. With R, you just rerun the code with the new input and get the updated figures and analyses! Later, we will learn how to incorporate code for figures along with statistical analysis into fully reproducible reports with output as `.html` (Website), `.pdf` (Adobe Acrobat), and `.docx` (Microsoft Word)

3.2 Graphical Concepts

Before we dive into coding for visualizations, there are a few universal graphing concepts that are important to understand in order to create publication-ready graphics in R: file formats, pixel dimension, screen vs print colours, and accessibility.

3.2 Vector vs. Raster

There are many different **file formats** that you can use to save individual graphs. Each format has a different suffix or extension in the file name like `.jpg`, `.png`, or `.pdf`. Once saved, you can send these to graphics programs for minor tweaks, or you can send them directly to academic journals for final publication.

Importantly, file formats for visualizations fall into two main classes: **Raster** and **Vector**.

Raster files save graphs in a 2-dimensional grid of data corresponding to pixel location and colour. Imagine breaking up your screen into a large `data.frame` object with each pixel represented by a cell, and the value of each cell holding information about the colour and intensity of the pixel. You are probably quite familiar with this 'pixel art' format if you've ever worked with a digital photo or played a retro video game made before 1993. Some popular Raster file types include *JPEG/JPG, PNG, TIFF, and BMP*.

Vector files save information about shapes. Instead of tracking every single pixel, the data are encoded as coloured points mapped onto a two-dimensional plane, with points connected by straight or curved lines. To generate the image, the computer must plot out all of the points and lines, and then translate that information to pixels that display on your screen.

If you've ever drawn a shape in a program like Microsoft Powerpoint or Adobe Illustrator, you might have some sense of how this works. Some popular vector formats include *SVG, PDF, EPS, AI, PS*.

So, why does this matter?

In most cases, you should save your visualizations as **vector** files. *SVG* is a good choice, because it can be interpreted by web browsers and it is not proprietary. *PDF* and *EPS* are proprietary, from the *Adobe* company. But they are commonly used by publishers and can be viewed on most computers after installing free software.

Saving your graphs as a *vector* format allows you (or the journal proof editor) to easily scale your image while maintaining crisp, clear lines. This is because the shapes themselves are tracked, so scaling just expands or contracts the x- and y-axes.

In contrast, if you expand a *raster* file, your computer has to figure out

3.2. GRAPHICAL CONCEPTS

how to expand each pixel. This introduces blurriness or other artifacts. You have seen some images that look 'pixelated' – this happens when you try to expand the size of a lower-resolution figure. This also happens when a program compresses an image to save space – the computer program is trying to reduce the data content by reducing the dimensionality of the image.

In summary, **vector** images are generally a better format to use when saving your figures because you can rescale to any size and the lines will always be clean and clear. There are a few important exceptions, however.

1. **Photographs** – Photographs captured by a camera are saved in the *Raster* format and cannot be converted to vector without significant loss of information.

2. **Grid Data** – Raster data are convenient for plotting data that occurs in a grid. This includes most spatial data that is broken up into a geographic grid. However, you may often want to use the vector format for mapping/GIS data so that the overlapping geographical features (e.g. borders, lakes, rivers) remain in the vector format, even if an underlying data layer is a raster object (e.g. global surface temperature in 1 × 1 km squares).

3. **Large Data** – With some large data applications (e.g. 'omics' datasets) a graphing data set may have many millions or even billions of data points or lines. In these cases, the **vector** file would be too big to use in publication (e.g. several gigabytes) or even too big to open on standard laptop or desktop computers. In this case you might opt for a high-resolution *Raster* file. On the other hand, you could graph your data using a representative subset of data or using a density grid with colours corresponding

to the density of points. In each case, you would keep the *vector* format to maintain clean lines for the graph axes and labels, even though some of the data is in the raster format.

The bottom line: you generally want to save your graphics as *SVG* or *PDF* files if you plan to publish them.

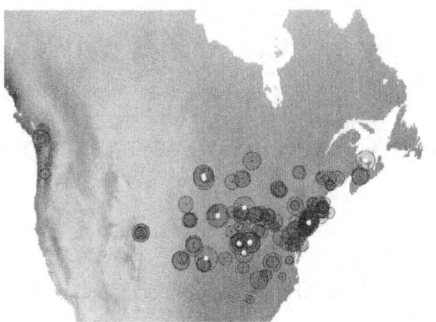 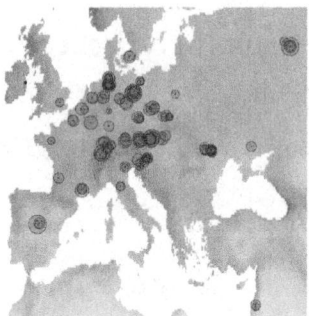

Figure 3.1: Raster map of temperature (blue) overlain with vector drawings of sampling locations (red), scaled by population size, from the *Global Garlic Mustard Field Survey*

3.2 Resolution vs Dimension

In cases where you do need to use raster images in a publication, pay careful attention to the image's **pixel dimension**. You have probably heard about image *resolution*: For example, a 2 megapixel camera is better than a 1 megapixel camera; a 200 dpi (dots per inch) printer is better than 50 dpi. But when creating and saving raster images, it's not just the **resolution** that matters, the image **size** also determines the quality of the final image. The **size** and **resolution** of an image jointly determine its **pixel dimension**.

> **Example**: An image with 200 dpi that is 1 x 3 inches will have the same pixel dimension of an image with 100 dpi that is 2

3.2. GRAPHICAL CONCEPTS

x 6 inches. These images will look exactly the same if they are printed at the same size. The pixel dimension, not the resolution *per se* determines how crisp or pixelated an image looks.

3.2 Screen vs Print

Another important consideration is the intended audience and whether they will likely view your figures on a computer screen or printed page (or both). Each pixel of your screen has tiny lights that determine the specific colour that is reproduced. The pixels *emit* different wavelengths from your screen, which overlap in our eyes to produce the different colours that we see. In contrast, printed images get their colour from combinations of ink, which *absorb* different wavelengths of colour. This is a key distinction! One important consequence of this difference is that your computer monitor can produce a broader range of colours and intensities than a printed page, and therefore some colours on your computer monitor can look very different in print. In print, the intensity of colours are limited by the intensity of the Cyan, Magenta and Yellow inks that are used to reproduce the images. This is called **CMYK printing**.

Some programs like *Adobe Illustrator* have options to limit the screen to display only those colours that can be reproduced with CMYK printing.

3.2 Accessibility

Another important consideration about your choice of colours involves your audience. Remember that many cultures have particular intuitions about colours that can cause confusion if your choice does not match these expectations. For example, in Western European cultures, the red spectrum colours (red, orange, yellow) are often associated with 'hot' or 'danger' while blue spectrum (blue, cyan, purple) are more associated with 'cold' or 'calm'. Given these biases, imagine how confusing it would be to look at a weather map that used blue for hot temperatures and red for cool temperatures.

In addition to **cultural biases**, a significant portion of the population has some form of **colour-blindness** that limits their ability to see certain colours. This article in Nature has a good explanation with tips for making inclusive figures: *https://www.nature.com/articles/d41586-021-02696-z*

The Simulated Colour Blind Palettes Figure shows a simulation of colour blindness, written in R. Note how certain reds and blue/purple are indistinguishable. A good strategy is to use different **intensity** as well as different **spectra**.

> **Note**: If you are reading a black-and-white version of this book, this link will take you to a colour version of the image:

https://github.com/ColauttiLab/RCrashCourse_Book/blob/master/images/colorblind.png

There is also a more practical reason for this. It is common for scientists and students to print your manuscript or published article in black-and-white to read on public transit or during a group discussion. If you choose colours and shading that can be interpreted properly in black and white, then you will avoid confusion with this significant portion of your audience. The `viridis` package is a good tool for this. See: *https://cran.r-project.org/web/packages/viridis/vignettes/intro-to-viridis.html*

3.3 Getting Started

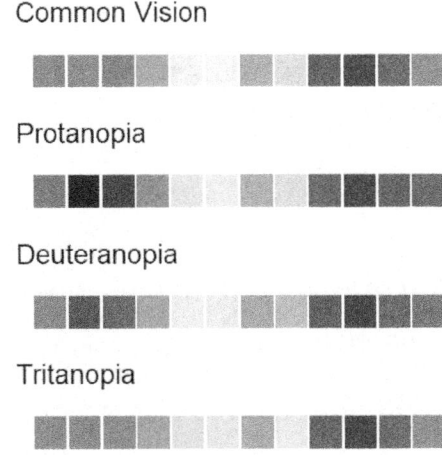

Figure 3.2: Simulated colourblind palettes

3.3 Getting Started

Install the ggplot2 package using the install.packages() function the first time you want to use it. This installs it on your local computer, so you only need to do it once – though it is a good idea to re-install periodically to update to the most recent version.

```
install.packages("ggplot2")
```

Once it is installed, you still need to load it with the library function if you want to use it in your code.

```
library(ggplot2)
```

3.3 Data setup

We will again be working with the FallopiaData.csv dataset, which can be downloaded here: *https://colauttilab.github.io/RCrashCourse/*

FallopiaData.csv

Save this text file to a folder called `Data` inside your project folder. Be sure to save it as `FallopiaData.csv` and make sure the `.csv` is included at the end of the name.

Remember that you can find your current working folder with the `getwd()` function. You may want to make a new R Project directory as discussed in the *R Fundamentals* chapter.

Now load the *.csv* file into R as a `data.frame` object:

```
MyData<-read.csv("./Data/FallopiaData.csv", header=T)
```

Alternatively, you can load the file right from the internet:

```
MyData<-read.csv(
  "https://colauttilab.github.io/RCrashCourse/FallopiaData.csv")
```

This dataset comes from the research group of Dr. Oliver Bossdorf at the University of Tübingen in Tübingen, Germany. It was published as part of a paper by Parepa and Bossdof *Testing for allelopathy in invasive plants: it all depends on the substrate!* (Biological Invasions, 2016), which you can find here: *https://doi.org/10.1007/s10530-016-1189-z*

> **Note**: Tübingen is a historic university in wonderful little town on the Neckar River. Let's inspect the `data.frame` that was created from the data, to see what kind of data we are working with.

```
str(MyData)
```

Figure 3.3: Punting on the Neckar river in Tübingen

```
'data.frame':    123 obs. of  13 variables:
 $ PotNum       : int  1 2 3 5 6 7 8 9 10 11 ...
 $ Scenario     : chr  "low" "low" "low" "low" ...
 $ Nutrients    : chr  "low" "low" "low" "low" ...
 $ Taxon        : chr  "japon" "japon" "japon" "japon" ...
 $ Symphytum    : num  9.81 8.64 2.65 1.44 9.15 ...
 $ Silene       : num  36.4 29.6 36 21.4 23.9 ...
 $ Urtica       : num  16.08 5.59 17.09 12.39 5.19 ...
 $ Geranium     : num  4.68 5.75 5.13 5.37 0 9.05 3.51 9.64 7.3 6.36 ...
 $ Geum         : num  0.12 0.55 0.09 0.31 0.17 0.97 0.4 0.01 0.47 0.33 ...
 $ All_Natives  : num  67 50.2 61 40.9 38.4 ...
 $ Fallopia     : num  0.01 0.04 0.09 0.77 3.4 0.54 2.05 0.26 0 0 ...
 $ Total        : num  67.1 50.2 61.1 41.7 41.8 ...
 $ Pct_Fallopia : num  0.01 0.08 0.15 1.85 8.13 1.12 3.7 0.61 0 0 ...
```

The data come from a plant competition experiment involving two invasive species from the genus *Fallopia*. These species were grown in planting pots in competition with several other species. The first four columns give information about the pot and treatments (Taxon = species of *Fallopia*. The rest give biomass measurements for each species. Each column name is the genus of a plant grown in the same pot.

3.4 Basic Graphs

Think back to the *R Fundamentals* Chapter, and you will hopefully recall the different data types represented by the columns of our data. For graphing purposes, there are really just two main types of data: **categorical** and **continuous**. Putting these together in different combinations with ggplot() gives us different default graph types.

Each ggplot() function requires two main components:

3.4. BASIC GRAPHS

1. The `ggplot()` function defines the input data structure. This usually includes a nested *aesthetic* function `aes()` to define the plotting variables and a `data=` parameter to define the input data.

2. the `geom_<name>()` function defines the output geometry

Note that the `<name>` denotes a variable name that differs depending on the geography used to map the data. A detailed example of a complex ggplot graph is covered in more detail in the next chapter. For now, we'll look at the most common visualizations, organized by input data type.

Specifically, we'll consider effective graphs for visualizing one or two variables, each of which may be categorical or continuous.

3.4 One Continuous

Usually when we only have a single continuous variable to graph, then we are interested in looking at the frequency distribution of values. This is called a **frequency histogram**. The frequency histogram shows the distribution of data – how many observations (y-axis) for a particular range of values or *bins* (x-axis).

It is very common to plot histograms for all of your variables before running any kind of statistical model to check for outliers and the distribution type. This is covered in the book *R STATS Crash Course for Biologists*. Looking at the histograms is a good way to look for potential coding errors (e.g. outliers) and whether data are generally normal or should be transformed to meet the assumptions of our statistical models.

Histogram

```
ggplot(aes(x=Total), data=MyData) + geom_histogram()
```

```
`stat_bin()` using `bins = 30`. Pick better value with
`binwidth`.
```

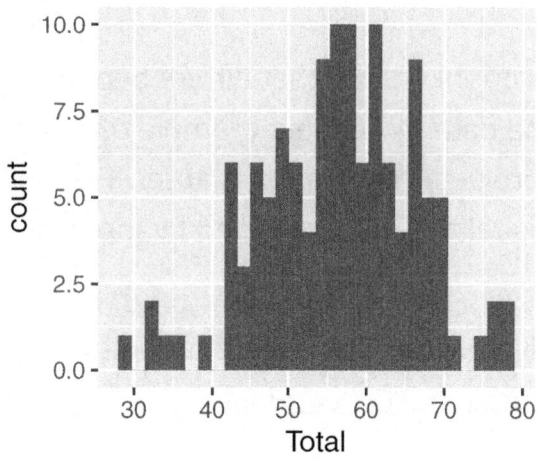

Note the two main components of our code: the `ggplot()` function defines the input data and the `geom_histogram()` function provides the mapping function.

Also note the warning message about `binwidth` in `stat_bin()`. This is not a problem, it is just R telling us that it chose a default value that may not be idea. We can try different values of `binwidth=` in the `geom_histogram()` function to specify the width of the bins along the x-axis. We'll look at this in more detail later.

You can also see we get a warning message about `stat_bin()`. With R we can distinguish *warning* messages from *error* messages.

Error messages represent bigger problems and generally occur when the function doesn't run at all. For example, if we wrote `total` instead of `Total`, we would get an error because R is looking for a column in `MyData` called `total`, which is not the same as `Total` with a capital *T*.

3.4. BASIC GRAPHS

Warning messages don't necessarily prevent the function from running, as in this case. However, it is still important to read the warning and understand if it is ok to ignore it. In this case, it is suggesting a different `binwidth` parameter. We'll come back to this later when we explore some of the different parameter options.

Density

A *density* geom is another way to graph the frequency distribution. Instead of bars, a smoothed curve is fit across the bins, and instead of 'count' data, an estimate of the probability is shown on the y-axis.

```
ggplot(aes(x=Total), data=MyData) + geom_density()
```

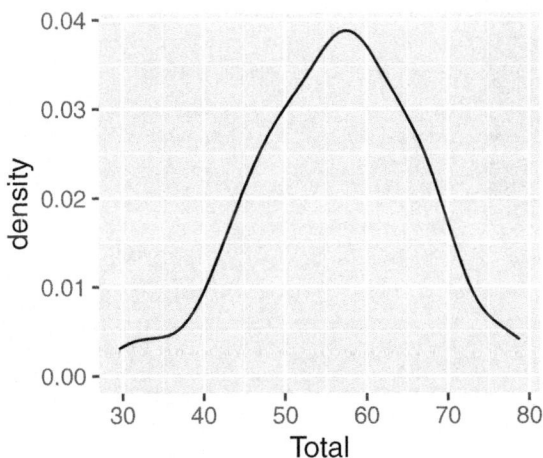

Notice that the y-axis says 'density', not probability? This curve is called a **probability density function** and we can calculate the probability of observing a value between two points along the x-axis by calculating the area under the curve for those two points. For example, integrating the total area under the curve should equal to a probability of 1.

3.4 One Categorical

If we input one variable that is *categorical* rather than *continuous*, then we are often most interested in looking at the sample size for each group in the category. In a classic ANOVA for example, you want to make sure you have a similar number of observations for each group.

Instead of geom_histogram() we use geom_bar()

Bar Graph

```
ggplot(aes(x=Scenario), data=MyData) + geom_bar()
```

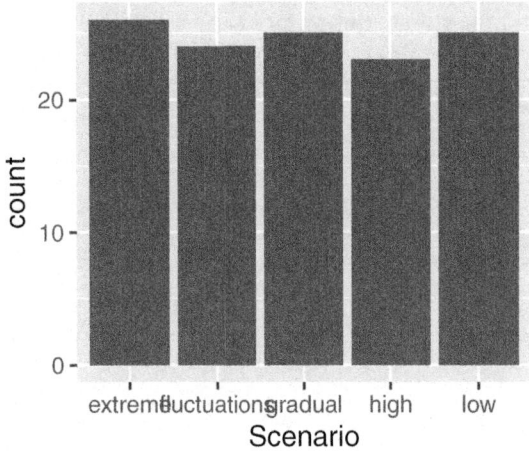

3.4 Two Continuous

If we have two continuous variables, we are most often interested in the classic **bivariate plot**, sometimes called a **scatter plot**.

The bivariate plot or scatter plot is the 'meat and potatoes' of data visualization. By plotting two variables we can see if they are independent ('shotgun' pattern) or have some degree of correlation (oval sloped up

3.4. BASIC GRAPHS

or down). We can also look for outliers, which would be seen as isolated points that are far away from the main 'cloud' of points.

```
ggplot(aes(x=Silene, y=Total), data=MyData) + geom_point()
```

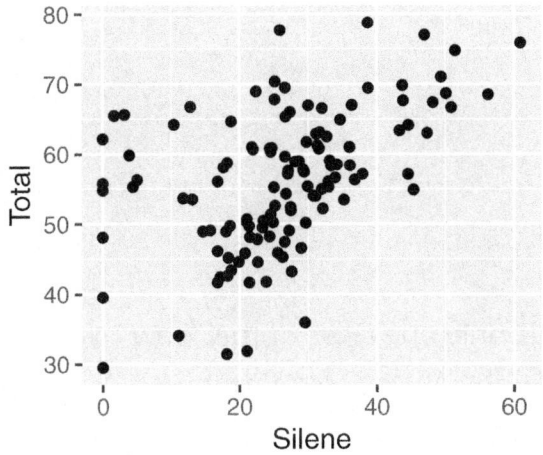

3.4 Two Categorical

Plotting two categorical variables is not so useful. Instead, we are better off making a summary table as described in detail the *Data Science Chapter*. Or, we can use the length function with aggregate() as introduced in the previous chapter:

```
aggregate(Total ~ Nutrients:Scenario, data=MyData, length)
```

	Nutrients	Scenario	Total
1	high	extreme	26
2	high	fluctuations	24
3	high	gradual	25
4	high	high	23
5	low	low	25

OR we can use the handy `table()` function if we want to summarize the variables as rows and columns:

`table(MyData$Scenario,MyData$Nutrients)`

```
              high low
extreme        26   0
fluctuations   24   0
gradual        25   0
high           23   0
low             0  25
```

In this case we can see that there is only one class of the "Low" *Nutrient* treatment, but four classes of "High" Nutrient treatments. In other words, all of the rows with "Low" Nutrient treatment also have the "Low" Scenario, and NONE of the rows with "High" Nutrient treatment have "Low" in the Scenario treatment. However, all groups have similar sample size of about 25. This is because the experiment compared low vs high nutrients, but also looked at different ways that high nutrients could be delivered.

3.4 Categorical by Continuous

If we have a categorical and a continuous variable, we usually want to see the distribution of points for the two variables. There are a few ways to do this:

3.4.5.1 Box plot

The box plot is a handy way to quickly inspect a few important characteristics of the data:

1. **median**: middle horizontal line (i.e. the 50th percentile)
2. **hinges**: top and bottom of the boxes showing the 75th and 25th percentiles, respectively
3. **whiskers**: vertical lines showing the highest and lowest values (excluding outliers, if present)
4. **outliers**: points showing outlier values more than 1.5 times the inter-quartile range (i.e. 1.5 times the distance from the 25th to 75th percentiles). Note that not all data sets will have outlier points.

```
ggplot(aes(x=Nutrients, y=Total), data=MyData) + geom_boxplot()
```

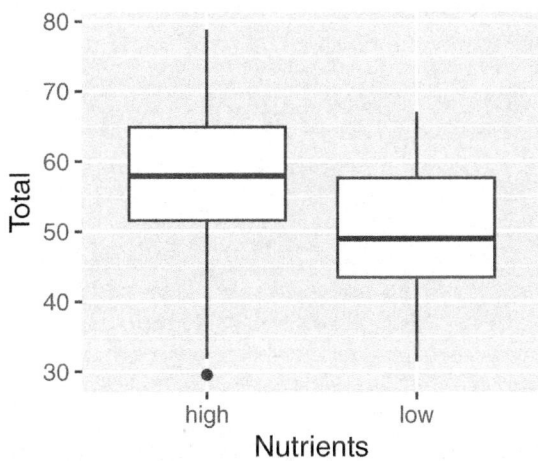

3.4.5.2 Violin plot

Violin plots or *Density strips* are another popular way to plot these type of data. The 'violin' or 'density' refers to the smoothed frequency distribution, which is similar the the geom_histogram we saw above, but imagine fitting a smoothed line along the top of each bar, and then turning it on its side and mirroring the image.

It can be very helpful to include both the violin and boxplot *geoms* on the same graph:

```
ggplot(aes(x=Nutrients, y=Total), data=MyData) +
  geom_violin() + geom_boxplot(width=0.2)
```

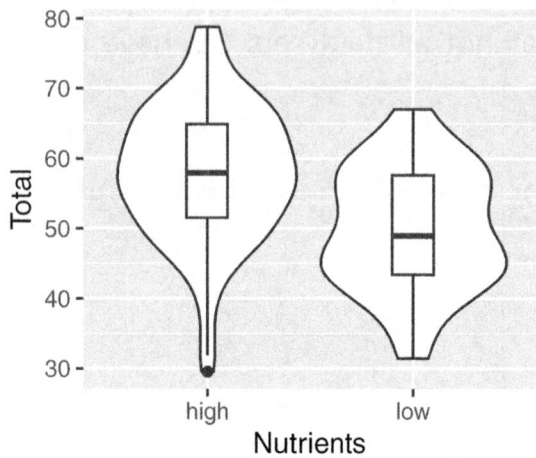

3.4.5.3 Dotplot

The dotplot stacks points of similar value. It's particularly useful for smaller datasets where the smooting of the density function may be unreliable. There are a couple of options here:

1. You can use dotplots for individual variables

```
ggplot(aes(x=Total), data=MyData) +
  geom_dotplot(binwidth=2)
```

3.4. BASIC GRAPHS

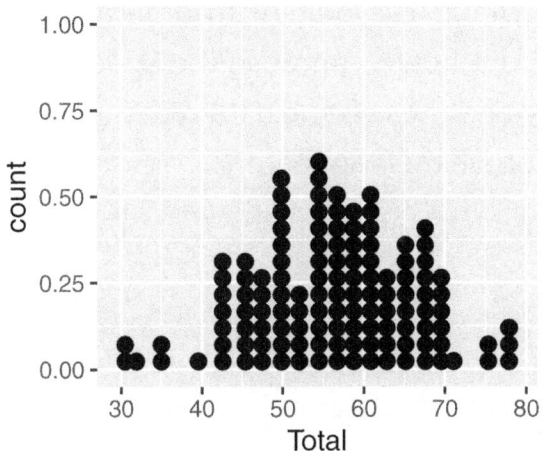

2. Or for continuous-by-categorical (but note the extra parameters)

```
ggplot(aes(x=Nutrients, y=Total), data=MyData) +
  geom_dotplot(binaxis="y", binwidth=2)
```

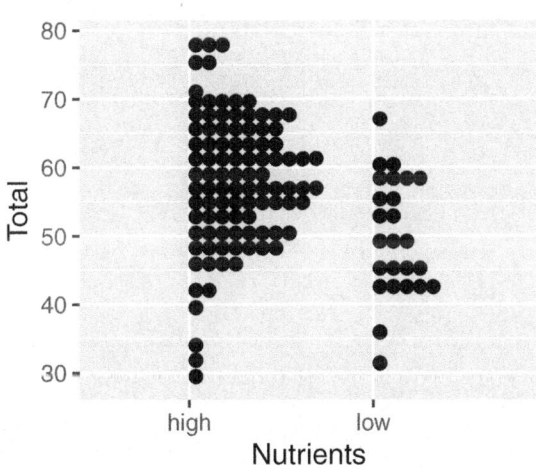

3.4 Done!

That's it! That's all you need to start exploring your data! Load your data frame, and plot different combinations of variables to look at the distribution of values or the relationship between your variables.

Once you are comfortable producing these plots with different data types, you might start thinking about how to improve the appearance of your graphs. Once you understand these basic data types, you can explore how to customize and improve their appearance.

Chapter 4

Basic Customization

4.1 Overview

Now that you know how to use R to to quickly generate graphs, we'll explore how to customize our graphs to make high-quality figures for publication.

There are a number of parameters and other functions available with `ggplot()` that you can use to quickly customize your graphs. First, we'll look at the different ways to customize the look and feel of our graphs. Then, we'll combine multiple graphs in the same multi-panel figure. In addition to generating more complex figures for publication, these multi-panel graphs can come in quite handy for exploring more complex data set.

4.2 Setup

Continuing from the last chapter, load the `ggplot2` library and set up our plotting data.

```
library(ggplot2)
MyData<-read.csv(
  "https://colauttilab.github.io/RCrashCourse/FallopiaData.csv")
```

4.3 `binwidth`

As we noted briefly in the last chapter, we can use `binwidth` with the histogram graph type to alter the size of the 'bins' along the x-axis. A bin is defined by a range of values (x-axis). The bin count or frequency (y-axis) shows the number observations (or fraction) that fall within each bin range.

The `binwidth` defines the range of values (i.e. width) of each bin. Here are a couple of examples for comparison.

```
library(ggplot2)
ggplot(aes(x=Total), data=MyData) + geom_histogram(binwidth=9)
```

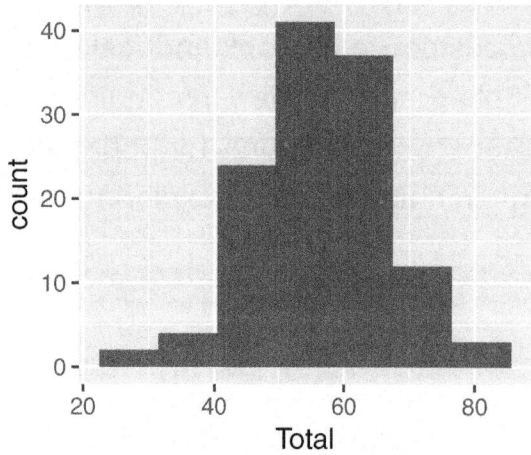

```
ggplot(aes(x=Total), data=MyData) + geom_histogram(binwidth=0.5)
```

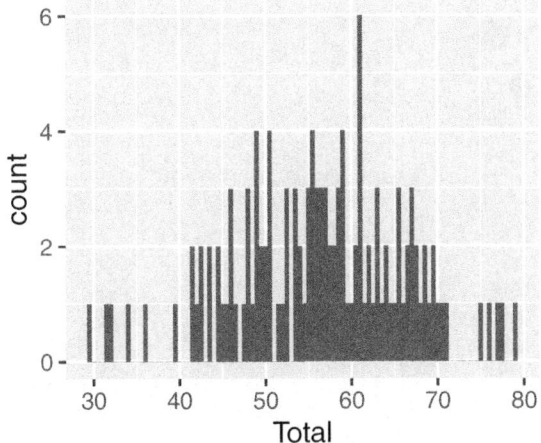

Compare the code for each graph to understand how binwidth affects *both* the y-axis values and the width of the blocks along the x-axis. Wider bins contain more observations, just like larger barrels catch more rain.

4.4 size

This controls the point size. Importantly, size values can be interpreted by R in two ways, which can cause some confusion:

1. As a single **value**: To assign a specific size to all points. This is assigned in the geom_point() function

```
ggplot(aes(x=Silene, y=Total), data=MyData) +
  geom_point(size=5)
```

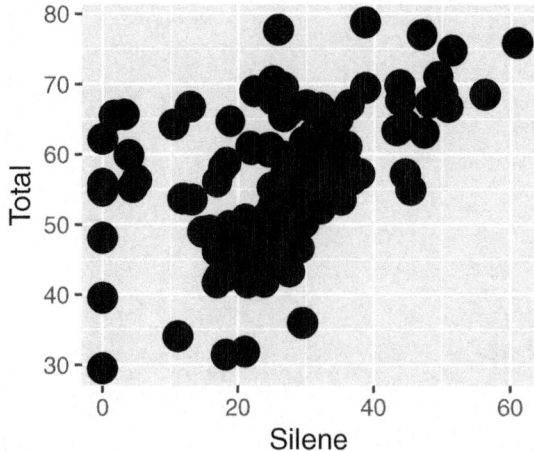

2. As a set of values defined in a **vector**: Scale size based on a column of data (e.g. number of observations). This is defined in the aes() function.

From the perspective of the R console, these are pretty much the same thing since a single value can be treated as a vector with just one element.

```
ggplot(aes(x=Silene, y=Fallopia), data=MyData) +
  geom_point(aes(size=Total))
```

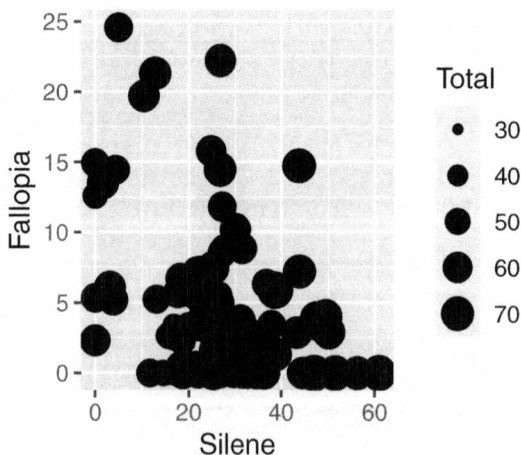

4.4. SIZE

NOTE: The following code will produce the exact same graph.

```
ggplot(aes(x=Silene, y=Fallopia, size=Total), data=MyData) +
  geom_point()
```

Compare this `ggplot()` function with the two previous.

> **Question**: What do you think is the difference between putting an aes function inside of `ggplot()` vs inside of `geom_point()`?

Answer: It's important to understand the difference, even though in this specific example it doesn't change the graph. Here is a short summary:

1. If we put a variable inside of `ggplot()` then the parameter applies to ALL of the *geom* functions that follow it.

2. If we put a variable inside of a *geom* like `geom_point()`, then the parameter applies ONLY to that specific geometric shape layer.

3. We use `aes()` when refereincing a column from our input data.

We'll dive into these ideas in more detail in the next chapter, when we start to produce more complicated graphs with multiple, overlapping *geoms*.

Before continuing, take a moment to make sure you understand the three different examples of code and resulting output above.

4.5 `alpha`

Think of `alpha` as a measure of opacity, ranging from 0 to 1 with 1 being the default – a solid point or line.

This is particularly useful for visualizing overlapping points.

```
ggplot(aes(x=Silene, y=Total), data=MyData) +
  geom_point(aes(colour=Nutrients), size=5, alpha=0.3)
```

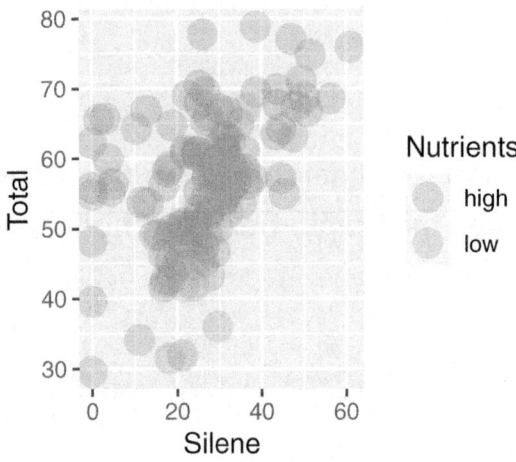

4.6 `colour (or color)`

Another nice feature of `ggplot` is that you can use alternate English spelling for some of the parameters. For example, you can use `colour=` or `color=` add colour to your color graphs.

Similar to point sizes, you can use colours in two main ways.

1. You can colour points based on a factor.

4.6. COLOUR (OR COLOR)

```
ggplot(aes(x=Silene, y=Fallopia), data=MyData) +
  geom_point(aes(colour=Nutrients))
```

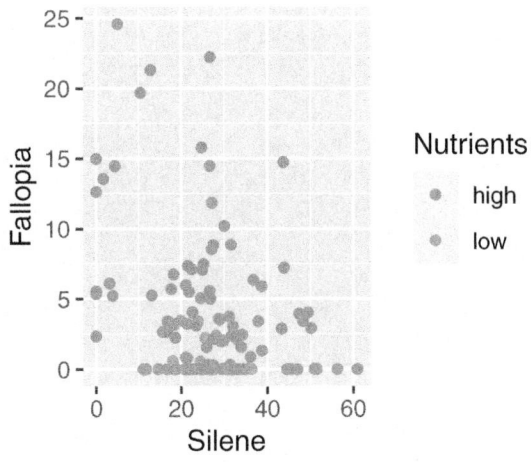

... or a continuous variable.

```
ggplot(aes(x=Silene, y=Fallopia), data=MyData) +
  geom_point(aes(colour=Total))
```

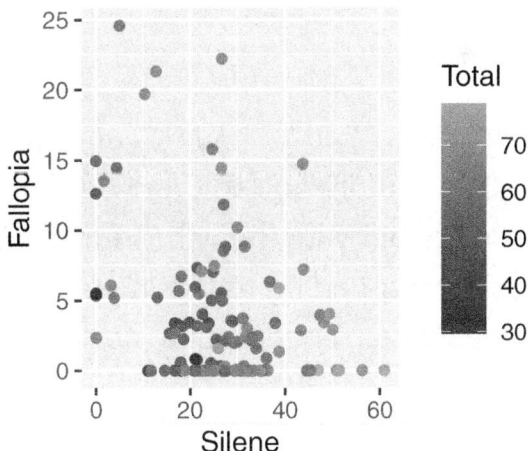

2. You can choose a specific colour to apply to all points.

```
ggplot(aes(x=Silene, y=Fallopia), data=MyData) +
  geom_point(colour="grey60")
```

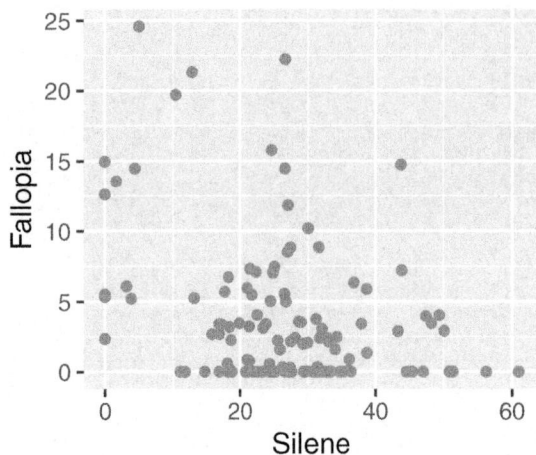

4.7 Colour with rgb()

Several colours are available as strings (e.g. "red", "blue", "aquamarine", "coral", "grey20", "grey60"), but if you can't find one that you want, you can make just about any colour with the rgb() function. The rgb function takes three values corresponding to the intensity of red, green and blue light, respectively. Values range from 0 (no colour) to 1 (brightest intensity).

```
ggplot(aes(x=Silene, y=Fallopia), data=MyData) +
  geom_point(colour=rgb(1,0.7,0.9))
```

4.7. COLOUR WITH RGB()

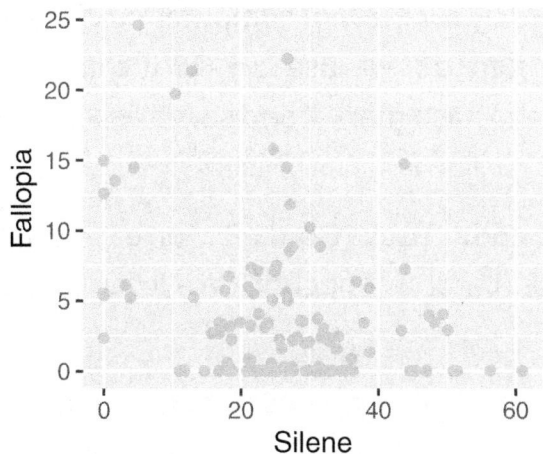

Some colouring systems use a 256-bit scale (0 to 255) instead of 0 to 1, which you can specify in the rgb() function with the maxColorValue = 255 parameter. See ?rgb for more information.

4.7 Hexadecimal Colour

Another common format for colour uses a **hexadecimal system**. In fact, the hexadecimal code is the output of the rgb() function that R uses for plotting:

rgb(0.1,0.3,1)

[1] "#1A4DFF"

I(rgb(255,0,0, maxColorValue=255))

[1] "#FF0000"

The **hexadecimal system** is a base-16 alphanumeric code that is common in computing. It uses the numerical digits 0-9 followed by the letters A (11) through F (16) as the 16 characters.

Hexadecimal colour codes are used by a variety of computer programs. For colouring visualizations with `ggplot`, we use a 6 OR 8-character **hexadecimal code**, starting with the hash mark # and saved as a string using quotation marks.

The **6-digit hexadecimal colour code** uses two digits for each base colour: red (r), green (g) and blue (b), or #<rrggbb>. We'll see an example to help clarify this.

This 6-digit code results in $16 \times 16 = 256$ shades of each colour, or $256^3 = 16,777,216$ total colour combinations

The **8-digit hexadecimal colour code** is similar, with the additional two digits at the end to define the level of alpha/transparency.

The `rgb()` function converts a vector of red, green, blue, (and optional alpha) to the 6- or 8-digit hexadecimal equivalent.

```
rgb(1,1,1,0.5)
```

```
[1] "#FFFFFF80"
```

Alternatively, transparency can be specified with the `alpha` parameter, as noted earlier.

4.7 Histogram

Note what happens when we use the `colour` parameter for a histogram.

```
ggplot(aes(x=Total), data=MyData) +
  geom_histogram(aes(colour=Nutrients), bins=10)
```

4.8. FILL

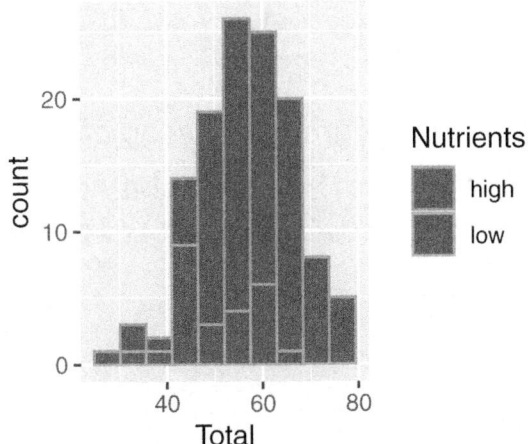

The coloured outlines might be useful in some cases, but we usually want the entire bars coloured. We can use the `fill` parameter for this.

4.8 fill

This parameter is used for histogram boxes and other geometric shapes that have a separate outline (`colour=`) and interior (`fill=`).

```
ggplot(aes(x=Total), data=MyData) +
  geom_histogram(aes(fill=Nutrients), bins=10)
```

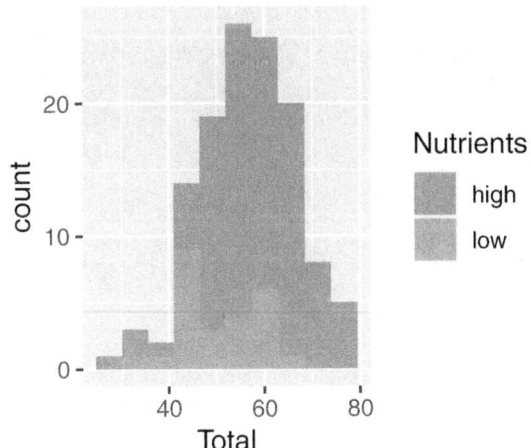

4.9 position

Use this to adjust the position, usually for histograms or bar graphs. For example, in the previous graph the bars are 'stacked' on top of each other. It can be hard to interpret a histogram with stacked bars, but we can shift the position using dodge.

```
ggplot(aes(x=Total), data=MyData) +
  geom_histogram(aes(fill=Nutrients), bins=10, position="dodge")
```

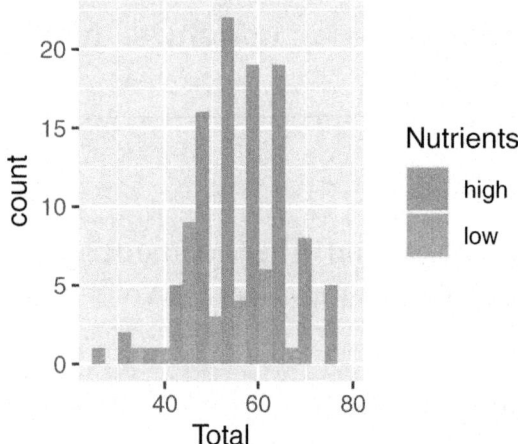

4.10 shape

You can also change the shape of your points, again using a column of data or a specific value.

```
ggplot(aes(x=Silene, y=Total), data=MyData) +
  geom_point(aes(shape=Nutrients))
```

4.10. SHAPE

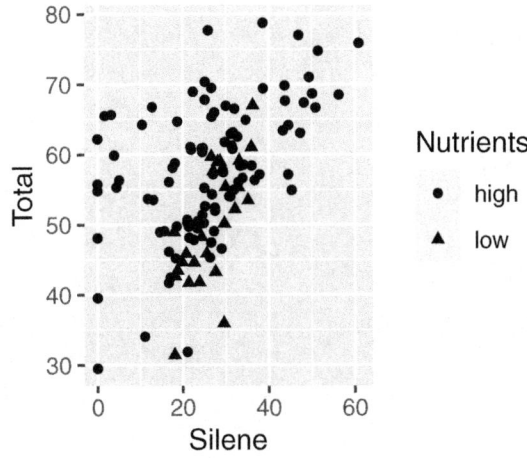

```
ggplot(aes(x=Silene, y=Total), data=MyData) +
  geom_point(shape=17)
```

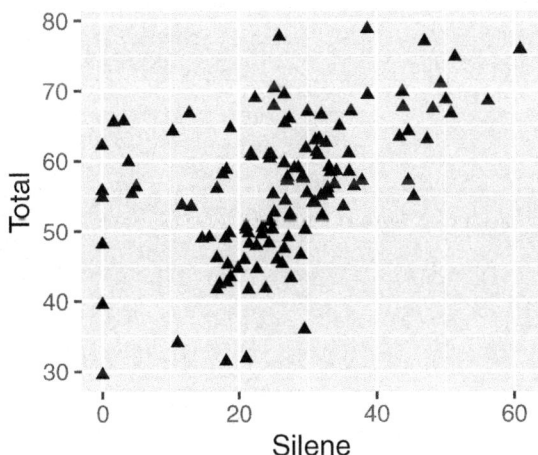

There are a number of different shapes available, by specifying a number from 0 through 25.

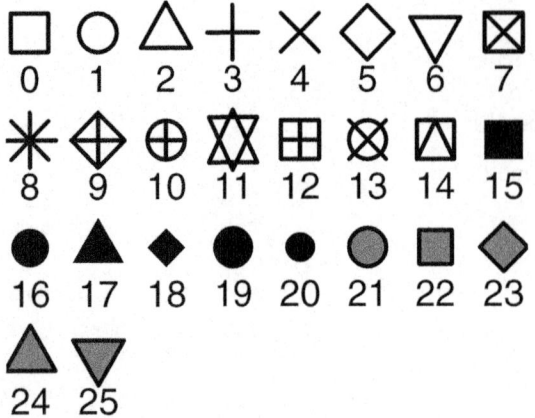

Note that the shapes with grey in the above figure can be coloured with `fill=` parameter, while all of the black parts (lines and fill) can be coloured with the `colour=` parameter.

You can use `fill` and `colour` to customize these separately.

```
ggplot(aes(x=Silene, y=Total), data=MyData) +
  geom_point(shape=21, size=5, colour="purple", fill="yellow")
```

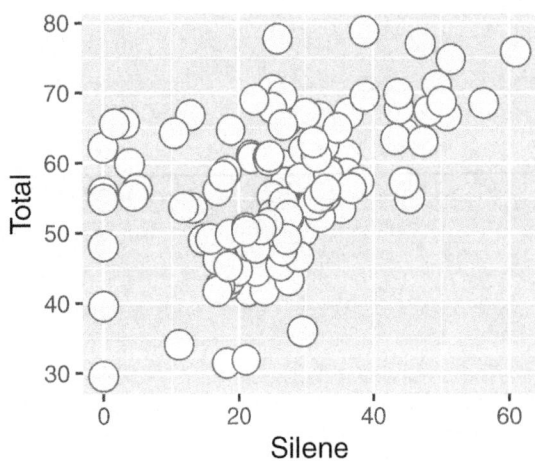

Note how a solid outline can help your points 'pop'.

Similarly, specifying a solid `colour` can definition to a histogram graph.

4.11. LAB, XLAB, AND YLAB

```
ggplot(aes(x=Silene), data=MyData) +
  geom_histogram(bins=20,colour="darkred",fill="aquamarine")
```

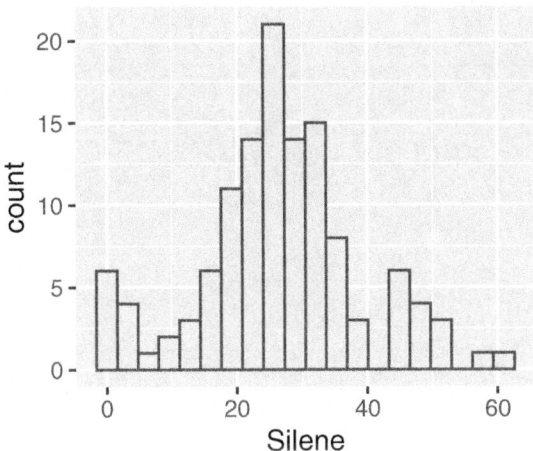

4.11 lab, xlab, and ylab

Use these to customize your axis labels.

```
ggplot(aes(x=Silene, y=Total), data=MyData) +
  geom_point() +
  xlab("Silene Biomass") + ylab("Total Biomass")
```

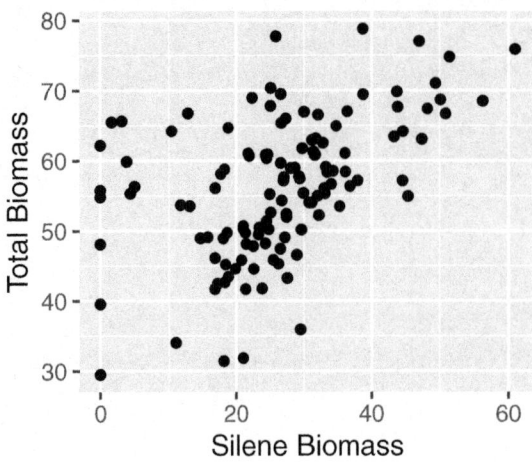

4.12 `labs`

This will add other labels to your plot. Usually you wouldn't use this for a figure intended for publication – for this you would need a detailed caption, usually just a paragraph of text below the figure. However, these can be useful for other documents: reports, websites, presentations, supplementary material, appendices, etc.

```
ggplot(aes(x=Silene, y=Total), data=MyData) +
  geom_point() + labs(title="Biomass", subtitle="More info here",
                caption="Appears after the figure")
```

4.13. THEMES AND GEOMS

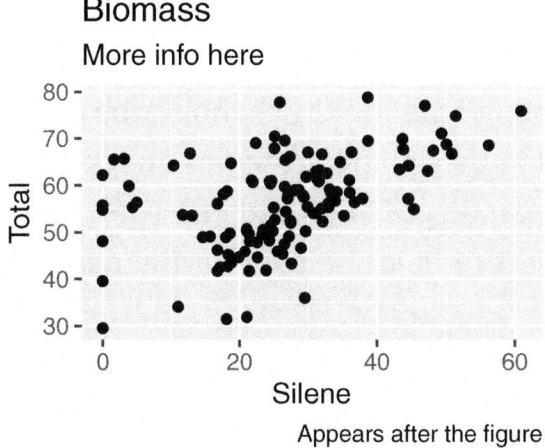

Appears after the figure

4.13 Themes and Geoms

We have already explored a few of the many **Geoms** available. These determine the *geometry* of your graph, which is how your data are mathematically mapped to the graphing space.

Themes define the look and 'feel' of your graphs.

In `ggplot()`, themes and geoms are added with a separate function linked to the graph by using the plus sign +.

4.13 `geom_<name>()`

We explored a few *geoms* above, but there are many more available on the `ggplot2` website, with helpful examples: https://ggplot2.tidyverse.org/reference/*

4.13 + theme_<name>()

There are a number of available themes, defined by changing the part of theme_<name>(). We'll try potting these diffent themes on the same graph. Rather than type out the same ggplot() and geom_ functions every time, we can define an object to hold the data for the plot, and then just change the theme.

4.13.2.1 Default theme:

```
Plot1<-ggplot(aes(x=Silene, y=Total), data=MyData) + geom_point()
Plot1 + theme_grey()
```

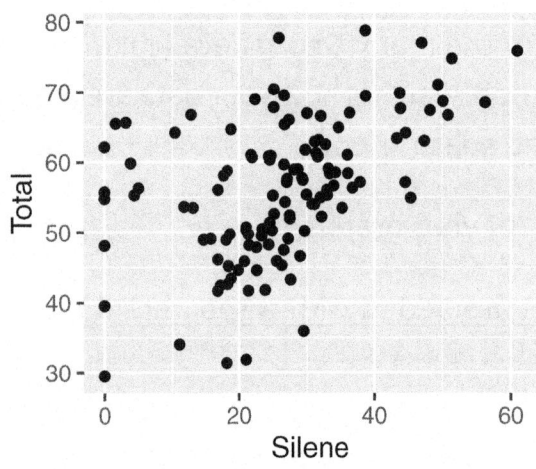

4.13.2.2 A cleaner theme with better contrast:

```
Plot1 + theme_bw()
```

4.13. THEMES AND GEOMS

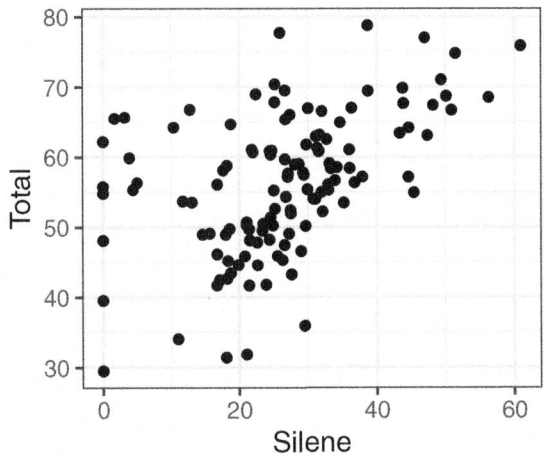

4.13.2.3 Thicker grid lines:

```
Plot1 + theme_linedraw()
```

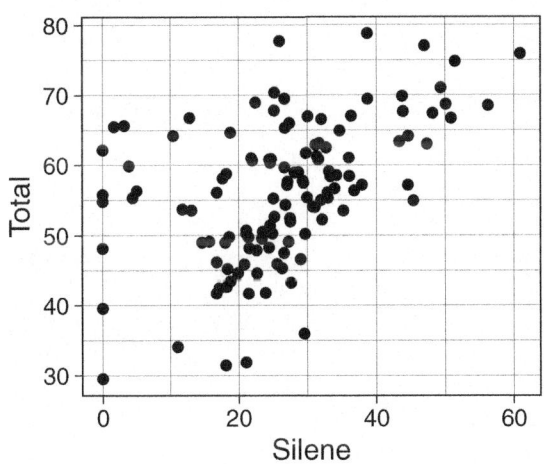

4.13.2.4 Fainter border and axis values

`Plot1 + theme_light()`

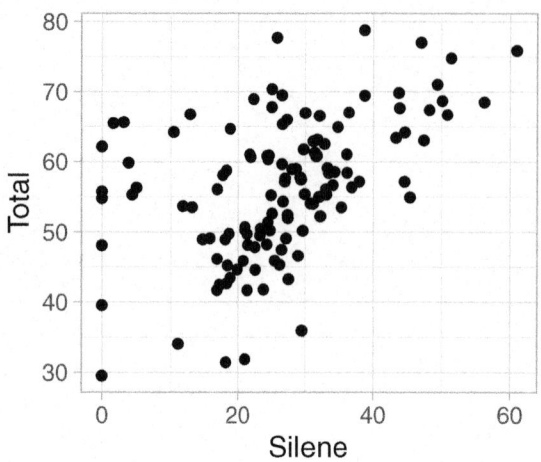

4.13.2.5 No borders at all

`Plot1 + theme_minimal()`

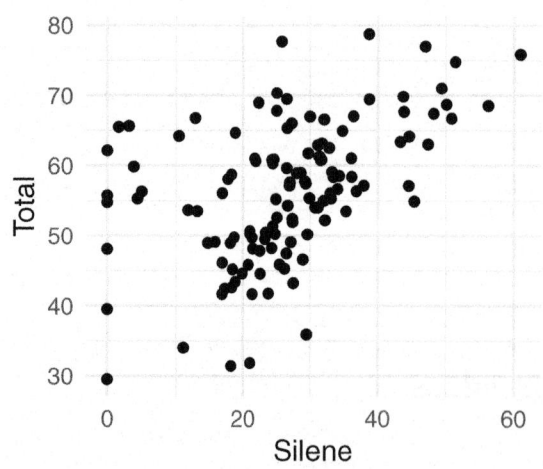

4.13. THEMES AND GEOMS

4.13.2.6 A minimal theme

This is closest to what you would see in a published paper, with x- and y-axis lines only

```
Plot1 + theme_classic()
```

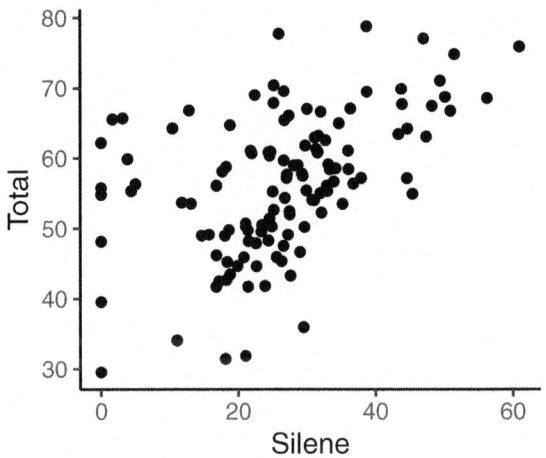

These can be further customized. Or you can create a completely new theme.

4.13 Custom Theme

Here is a simplified and cleaner version of `theme_classic` but with bigger axis labels that are more suitable for figures in presentation or publication. The theme is a function, which can be customized. Custom functions are covered in the Advanced R Chapter. For now you can just copy the code block below.

```
# Clean theme for presentations & publications
theme_pub <- function (base_size = 12, base_family = "") {
  theme_classic(base_size = base_size,
```

```
                   base_family = base_family) %+replace%
    theme(
      axis.text = element_text(colour = "black"),
      axis.title.x = element_text(size=18),
      axis.text.x = element_text(size=12),
      axis.title.y = element_text(size=18,angle=90),
      axis.text.y = element_text(size=12),
      axis.ticks = element_blank(),
      panel.background = element_rect(fill="white"),
      panel.border = element_blank(),
      plot.title=element_text(face="bold", size=24),
      legend.position="none"
    )
}
```

To use this theme, you have to make sure you run the entire function (e.g. highlight every line and press `Ctl + R` or click Run in *R Studio*).

Alternatively, you could *save it* as a separate .R file (e.g. `theme.R`) and then load it with the `source()` function (e.g. `source("./theme.R")`)

4.13 Publication Theme

A third, even easier option, is to load the version of this code that is available online.

```
source("http://bit.ly/theme_pub")
```

The theme is called `theme_pub` (pub is short for publication). To use it, run the above line, and then add it to your graphing functions:

4.13. THEMES AND GEOMS

```
Plot1 + theme_pub()
```

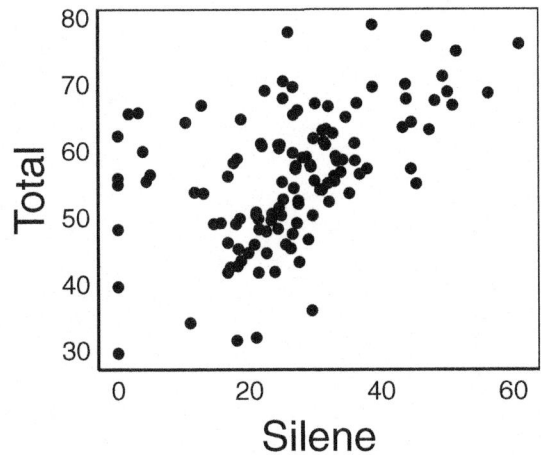

```
ggplot(aes(x=Silene),data=MyData) +
  geom_histogram(binwidth=2) + theme_pub()
```

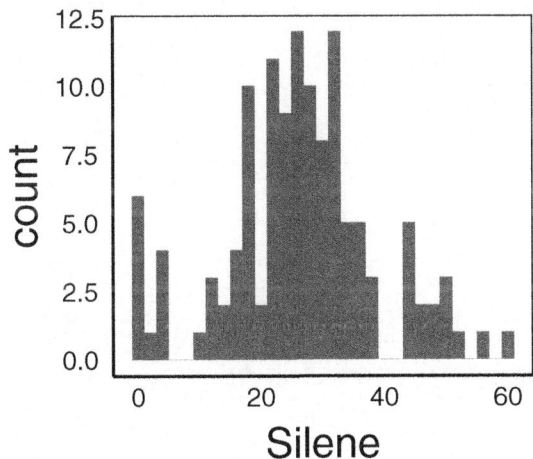

4.13 theme_set

If you want to use the same theme throughout your code, you can use the theme_set function.

```
theme_set(theme_pub())
Plot1
```

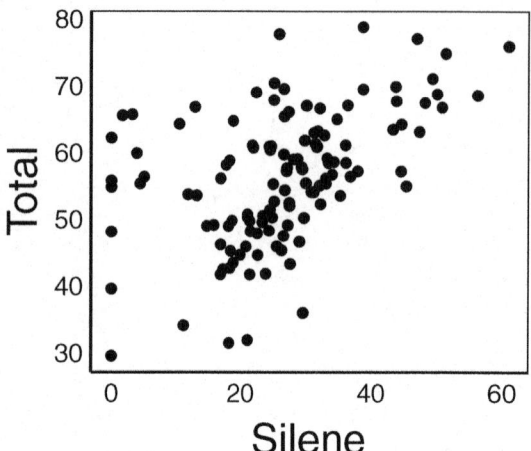

Now that we have run the `source` and `theme_set` functions, all of the graphs we make in this session will use the improved formatting. No more ugly grey background and tiny axis labels!

4.14 Basic Multi-Plot Graphs

It is often handy to plot separate graphs for different categories of a grouping variable. This can be done with `facets` in `qplot`.

4.14 `facets`

Facets have the general form VERTICAL ~ HORIZONTAL. Note the use of the tilde (~), not the dash (-). Use a period (.) to indicate 'all data' or 'do not separate my data', as shown in the following examples.

4.14.1.1 Vertical stacking

```
Plot2<-ggplot(aes(x=Silene),data=MyData) +
  geom_histogram(binwidth=2)

Plot2 + facet_grid(Nutrients~.)
```

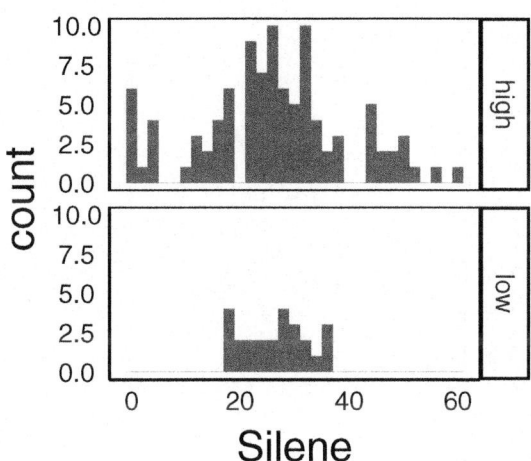

4.14.1.2 Horizontal stacking

```
Plot2 + facet_wrap(.~Nutrients)
```

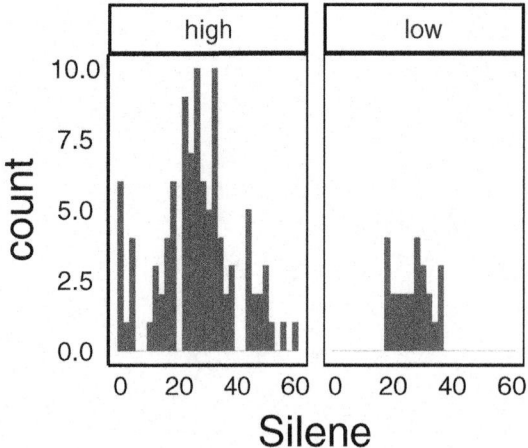

4.14.1.3 Horizontal by Vertical

```
Plot2 + facet_grid(Taxon~Nutrients)
```

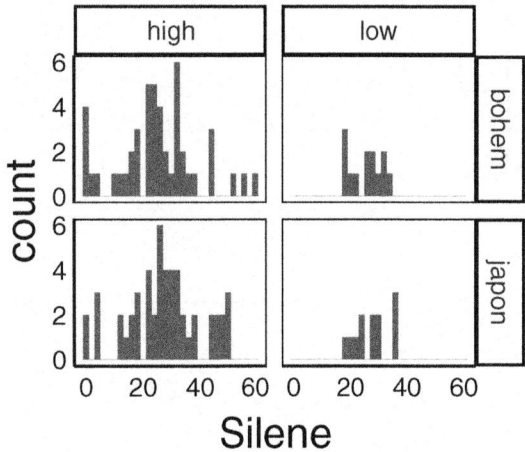

4.15 Graph output

Graphing in R studio is okay for exploration but eventually you are going to want to save those beautiful figures you made, and this can be part

4.15. GRAPH OUTPUT

of your reproducible workflow.

Writing code in R to save your graphs to an external file requires three important steps:

1. Open a file using a function like pdf or svg for the **vector** format, or png for the **raster** format. Remember that you usually will want to stick with a vector format, for reasons discussed in the *Graphical Concepts* section earlier.
2. Run the code to produce the graph. Instead of seeing a graph in your R interface, you will not see anything because the graph is being sent to the file.
3. **IMPORTANT**: Close the file! Do this with the dev.off() function.

Failing to close the file is a common source of error when saving graphs. If you are having problems with graphing outputs, try running the dev.off() function a few times to make sure you close any files that are 'hanging' open.

Here's an example code for making a pdf output of a graph. When you run it you should see a file appear in your working folder (you may have to refresh).

```
pdf("SileneHist.pdf") # 1. Open
  Plot2 + facet_grid(Taxon~Nutrients) # 2. Write
dev.off() # 3. Close
```

Note how the plotting function on the second line does not open in the plots window when you run this. This is because the info is sent to *SileneHist.pdf* file instead of the graphing area in *R Studio*.

4.16 Practice

Graphing may seem slow and tedious at first, but the more you practice, the faster you will be able to produce meaningful visualizations.

Don't be afraid to try new things. Try mixing up components and see what happens. At worst you will just get an error message.

Once you have a good understanding of these basics, you can see how to build more advanced plots in the next chapter.

Chapter 5

Advanced Visualizations

5.1 Overview

Before continuing with this tutorial/chapter, you should be familiar with the basics of ggplot from the previous chapters, and you should have lots of practice making graphs with different formatting options.

In this chapter, we look at some important *rules of thumb* for making professional and effective visualizations. Then, we will work through a detailed example with more advanced options for visualizations with the `ggplot` function. This includes everything you will need to make professional-grade figures.

The `ggplot` *cheat sheet* may be useful for you, both for this chapter and in the future when making your own figures. A *cheat sheet* is a printable file that provides a good summary and quick-reference guide for a particular function or activity. In this case, the `ggplot` *cheat sheet* provides a quick summary of a lot of the main graph types and parameters in the `ggplot2` library. It can be found along with other useful cheat sheets on the *Posit* website: *https://posit.co/resources/cheatsheets/*

5.2 Getting Started

First, we'll load the `ggplot2` library and set a custom plotting theme as described in the previous chapter.

```
library(ggplot2)
source("http://bit.ly/theme_pub")
theme_set(theme_pub())
```

The `source` function loads an external file, in this case from the internet. The file is just a script saved as .R file with a custom function defining different aspects of the graph (e.g. text size, line width, etc.) You can open the link in a web browser or download and open in a text editor to see the script

The `theme_set()` function sets our custom theme (`theme_pub`) as the default plotting theme. Since the theme is a function in R, we need to include the brackets, even though twe don't want to change anything in the function: `theme_pub()`

5.3 Rules of thumb

Before we dig into the code, it's worth reviewing some more general graphical concepts. Standards of practice for published graphs in professional journals can vary depending on format (e.g. print vs online), audience, and historical precedent. Nevertheless, there are a number of useful 'rules of thumb' to keep in mind. These are not hard and fast rules but helpful for new researchers who aren't sure how or where to start. In making decisions, always think of your audience and remember that the main goal is to communicate information as clearly and efficientl as possible.

5.3. RULES OF THUMB

1. Minimize 'ink'

In the old days, when most papers were actually printed and mailed to journal subscribers, black ink was expensive and printing in colour was very expensive. Printing is still expensive but of course most research articles are available online where there is no additional cost for colour or extra ink. However, the concept of minimizing ink (or pixels) can go a long way toward keeping a graph free from clutter and unnecessary distraction.

2. Use space wisely

Empty space is not necessarily bad, but ask yourself if it is necessary and what you want the reader to take away. Consider the next two graphs:

```
Attaching package: 'dplyr'
```

```
The following objects are masked from 'package:stats':

    filter, lag
```

```
The following objects are masked from 'package:base':

    intersect, setdiff, setequal, union
```

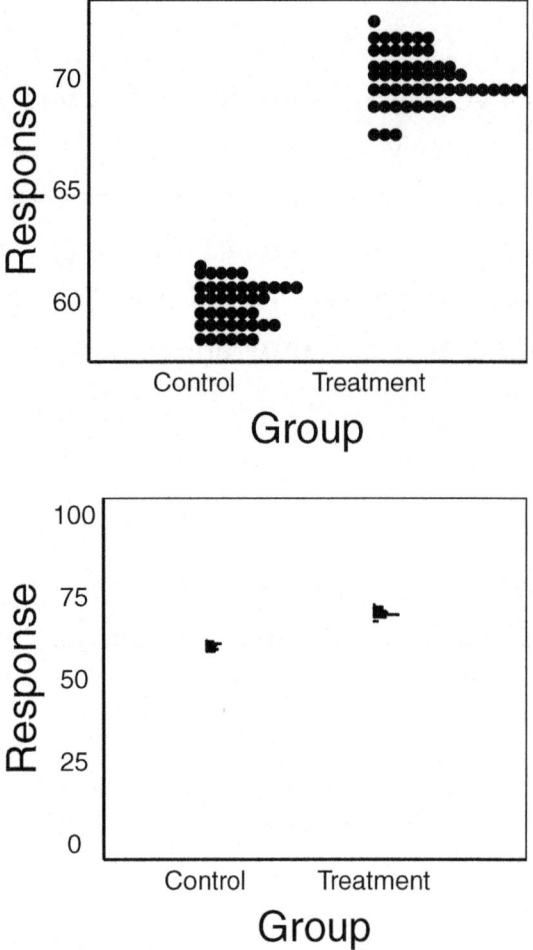

In the first example, the Y-axis is scaled to the data. In the second case, Y-axis scaled between 0 and 100.

> **Question**: What are the benefits/drawbacks of scaling the axes? When might you choose to use one over the other?

3. Choose a colour palette

Colour has three basic components:

a. **Hue** – the relative proportion of red vs green vs blue light

b. **Saturation** – how vivid the colour is

c. **Brightness** – the amount of white (vs black) in the colour

The abbreviation HSB is often used, or HSL (L = 'Lightness') or HSV (V = 'Value').

In R these can be easily defined with the rgb() function. For example:

- rgb(1,0,0) – a saturated red
- rgb(0.1,0,0) – a dark red (low brightness, low saturation)
- rgb(1,0.9,0.9) – a light red (high brightness, low saturation)

Don't underestimate the impact of choosing a good colour palette, especially for presentations. Colour theory can get a bit overwhelming but here are a few good websites to help:

- Quickly generate your own palette using Coolors: *https://coolors.co*
- Use a colour wheel to find complementary colours using Adobe: *https://color.adobe.com/create*

4. Colours have meaning

Try running this code and veiw the output in colour:

```
X<-rnorm(100)
Y<-X+seq_along(X)
D<-data.frame(Temperature=Y,Location=X,Temp=Y/3)
qplot(Location, Temperature,colour=Temp, data=D) +
  scale_color_gradient(high="blue", low="red")
```

Question: What strikes you as odd about this graph (not shown)?

Technically, there is nothing wrong. But we naturally associate colours with particular feelings. In this case, intuitively we associate red with hot and blue with cold, which is the opposite of what is shown in this graph. Be mindful of these associations when choosing a colour palette.

Another important consideration is that not everyone sees colour the same way. About 5% to 10% of the population has colour blindness. In order to make colour graphs readable to everyone, you should make sure to use colours that can still be interpreted when printed in greyscale, as explained in the Quick Visualizations Chapter.

5. Maximize contrast

Colours that are too similar will be hard to distinguish.

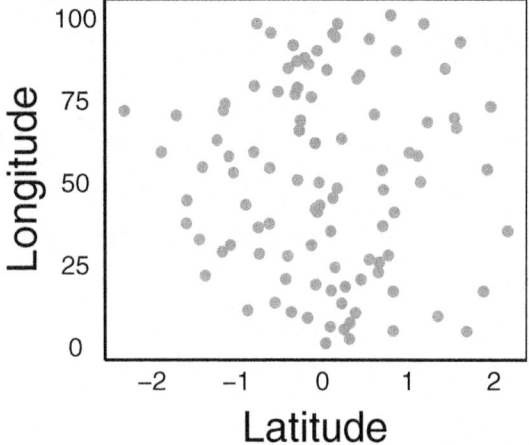

Can you see the gradient of colours? The difference among colours is called contrast, and generally a high-contrast palette is more informative than a low-contrast palette. Here is a plot of the same data, plotted with a wider range of colours:

```
ggplot(aes(Latitude,Longitude,colour=Precip), data=D) +
  geom_point() +
  scale_color_gradient(high="cyan", low="red")
```

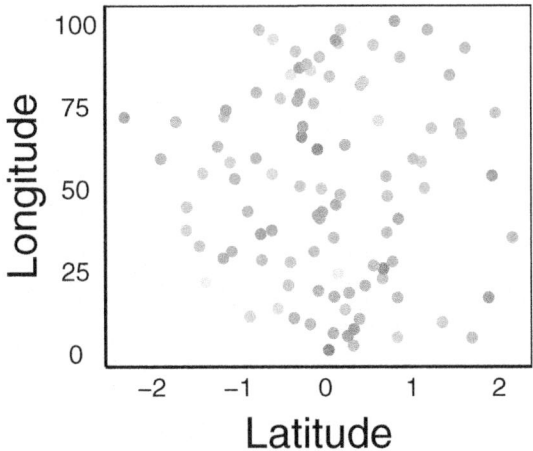

6. Keep relevant information

Make sure to include proper axis **labels** (i.e. names) and **tick marks** (i.e. numbers or categories showing the different values). These labels, along with the figure caption, should act as a stand-alone unit. The reader should be able to understand the figure without having to read through the rest of the paper.

7. Choose the right graph

Often the same data can be presented in different ways but some are easier to interpret than others. Think carefully about the story you want to present and the main ideas you want your reader to get from your figures. Look at these two graphs that show the same data.

```
ADat<-data.frame(Biomass=rnorm(100), Treatment="Treatment A")
BDat<-data.frame(Biomass=5 + rnorm(100) *5 +
                 ADat$Biomass * 5, Treatment="Treatment B")
PDat<-rbind(ADat,BDat)
ggplot(aes(Biomass, fill=Treatment), data=PDat)+
  geom_histogram(posit="dodge")
```

`stat_bin()` using `bins = 30`. Pick better value with

`binwidth`.

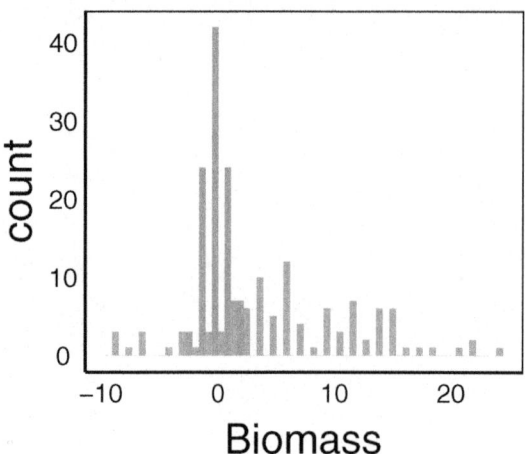

The first graph tells a story about the distributions – the mean and variance of each treatment

```
ggplot() + geom_point(aes(ADat$Biomass,BDat$Biomass)) +
   xlab("Biomass in Treatment A") + ylab ("Biomass in Treatment B")
```

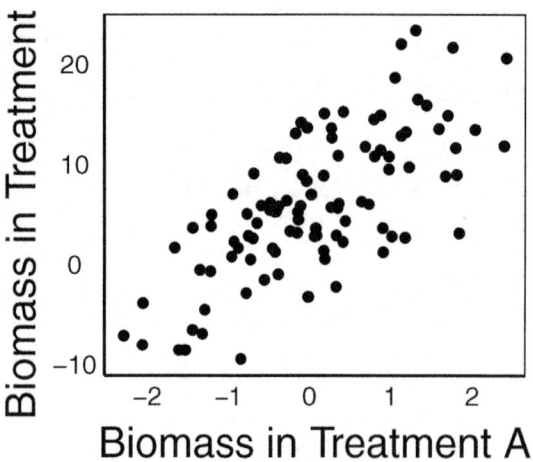

The second graph tells a story about the correlated relationship between Treatment A and Treatment B.

One graph is not necessarily better than the other. It depends on the story you want to tell.

5.4 Example

Now that we have gone over some basic graphing concepts, let's look at how to build a professional-grade figure. In fact, we'll reconstruct a figure published in a paper by Colautti & Lau in the journal Molecular Ecology (2015): *https://doi.org/10.1111/mec.13162*

5.4 Setup

The paper is a meta-analysis and review of evolution occurring during biological invasions. We will recreate Figure 2, which shows the result of a meta-analysis of selection gradients (β) and selection differentials (s). First, we'll just recreate the top panel, and then we'll look at ways to make more advanced multi-panel graphs like this.

The data from the paper are archived on Dryad: *https://datadryad.org/stash/dataset/doi:10.5061/dryad.gt678*

You could download the zip file and look for the file called `Selection_Data.csv` and save it to your working directory. But I have also put it on *Github*, so that you can download it directly to R:

```
SelData<-read.csv(
  "https://colauttilab.github.io/RCrashCourse/Selection_Data.csv")
```

We are also going to change the column names in the file to make them a bit more intuitive and easier to work with in R.

```
names(SelData)<-c("Collector", "Author", "Year", "Journal",
                  "Vol", "Species", "Native", "N",
                  "Fitness.measure", "Trait", "s",
                  "s.SE", "s.P", "B", "B.SE", "B.P")
```

5.4 Inspect

Let's take a quick look at the data

```
head(SelData)
```

```
          Collector              Author Year
1 KingsolverDiamond Alatalo and Lundberg 1986
2 KingsolverDiamond Alatalo and Lundberg 1986
3 KingsolverDiamond Alatalo and Lundberg 1986
4 KingsolverDiamond       Alatalo et al. 1990
5 KingsolverDiamond       Alatalo et al. 1990
6 KingsolverDiamond       Alatalo et al. 1990
              Journal            Vol             Species
1           Evolution      40:574-583   Ficedula hypoleuca
2           Evolution      40:574-583   Ficedula hypoleuca
3           Evolution      40:574-583   Ficedula hypoleuca
4 American Naturalist 135(3):464-471  Ficedula albicollis
5 American Naturalist 135(3):464-471  Ficedula albicollis
6 American Naturalist 135(3):464-471  Ficedula albicollis
  Native    N       Fitness.measure         Trait     s
1    yes  641   male mating success tarsus length -0.01
2    yes  713 female mating success tarsus length  0.01
3    yes 1705              survival tarsus length  0.04
4    yes <NA>              survival  tarus length  0.02
5    yes <NA>              survival  tarus length  0.08
6    yes <NA>              survival  tarus length  0.19
  s.SE s.P    B B.SE B.P
1        ns   NA
```

```
2          sig   NA
3          ns    NA
4          ns  -0.06
5          ns  -0.01
6          sig  0.01
```

It's worth taking some time to look at this to understand how to encode data for a meta-analysis. The **collector** column indicates the paper that the data came from. The **Author** indicates the author(s) of the original paper that reported the data. The **Year, Journal**, and **Vol** give information about the publication that the data came from originally.

We can see above the collector KingsolverDiamond, which represents a paper from Kingsolver and Diamond that was itself a meta-analysis of natural selection. Most of the studies came from this meta-analysis, but a few of the more recent papers were added by grad students, denoted by initials:

```
unique(SelData$Collector)
```

```
[1] "KingsolverDiamond"  "JAL"
[3] "DJW"                "CPT"
```

Species is the study species, and **Native** is its status as a binary yes/no variable. **N** is the sample size and **Fitness.measure** is the specific trait that defines fitness. **Trait** is the trait on which selection was measured. Finally, s is the **selection differential** and β is the **selection gradient**. Note that these are slopes in units of relative fitness per trait standard deviation. This is explained in more detail below.

5.4 Absolute Value

In this analysis, we are interested in the magnitude but not the direction of natural selection. In other words we would want to treat a slope of -4 the same as a slope of +4 because they have the same magnitude. Therefore, we can replace the s column with $|s|$

```
SelData$s<-abs(SelData$s)
```

We'll also add a couple of columns with random variables that we can use later to explore additional plotting options.

First, a column of values sampled from a z-distribution – this is a Gaussian (a.k.a. 'normal') distribution with mean = 0 and sd = 1.

```
SelData$Rpoint<-rnorm(nrow(SelData))
```

Second, a columnn of 1 and 0 sampled randomly with equal frequency ($p = 0.5$)

```
SelData$Rgroup<-sample(c(0,1), nrow(SelData), replace=T)
```

> **Question**: Do you remember `rnorm()` and `sample()` from the *R Fundamentals* Chapter?

If not, it may be a good time for a quick review. Remember to keep practicing – recognizing code is not the same as being able to write it from scratch.

5.4 Missing values

Before we plot the selection data, we should take a quick look at the values to check for potential errors.

```
print(SelData$s)
```

The output isn't shown here, but note that NA is used to denote missing data in the output.

We can subset to remove missing data:

```
SelData<-SelData[!is.na(SelData$s),]
```

Recall from the *R Fundamentals* Chapter that ! means 'not' or 'invert'

There is also a convenient drop_na function in the tidyr package:

```
library(tidyr)
SelData<-SelData %>%
  drop_na(s)
```

5.5 Measuring Selection

5.5 Don't Panic!

We're going to get a bit technical here. Don't worry if you don't completely understand all of the stuff below about measuring selection. Keep it for reference in case you decide you want to use it for your own research. For now, just try to understand it as well as you can and focus on the code used to produce the figures.

An analysis of phenotypic selection was proposed by Lande & Arnold (1983) as a simple but powerful tool for measuring natural selection. It is just a linear model with **relative fitness** on the y-axis and the **standardized trait value(s)** on the x-axis.

5.5 Relative Fitness

Fitness can be measured in many ways, such as survival or lifetime seed or egg production. Check out the list of specific fitness measures used in these studies:

```
unique(SelData$Fitness.measure)
```

(output not shown)

Absolute fitness is just the observed value (e.g. seed set or survival yes/no). Technicallly, we call these *fitness components* because they are not fitness *per se*, but they represent measurements of survival and reproduction, which are the key components that jointly determine fitness.

Relative fitness is just the *absolute* fitness divided by the mean. **Absolute fitness** is usually denoted by the capital letter W and **relative fitness** is usually represented by a lower-case w or omega ω. Expressing this in mathematical terms:

$$\omega = W_i / \bar{W}$$

where W_i is the mean of the study sample.

5.5 Trait Value

A **Trait Value** is just the measured trait on which selection may act. Use unique(SelData$Trait) to see the list of specific traits that were measured in these studies. The **Standardized Trait Value** is the traits z-score. See the *Distributions* Chapter in the book *R STATS Crash Course* for more information about z-scores. To calculate the z-score, we take each value, subtract the mean, and then divide by the standard deviation of the sample (sd):

$$\frac{x_i - \bar{X}}{sd}$$

Since traits have different metrics, they are hard to compare: e.g. days to flower, egg biomass, foraging intensity, aggression. But standardizing traits to z-scores puts them all on the same scale for comparison. Specifically, the scale of selection will be in standard deviations from the mean.

5.5 s vs β

Selection differentials (s) and selection gradients (β) measure selection using linear models but represent slightly different measurements. Linear models are covered in the *Linear Models* Chapter in the book *R STATS Crash Course*.

Both models use relative fitness (ω) as the response variable.

Selection differentials (s) measure selection on only a single trait, ignoring all other traits. In theory, the response to selection is a simple function of the genetic correlation between a trait and fitness.

Fitness differences among individuals can depend on a lot of things – genetic variation for the trait itself, but also environmental effects on the trait as well as effects on other traits that are under selection and correlated with the trait of interest.

Selection gradients (β) measure selection on a trait of interest while also accounting for selection on other correlated traits. This is done via a **multiple regression** – a linear model with multiple predictors.

Now that we have reviewed the relevant biological background, we can plot s and β to compare their distributions.

5.6 Distribution Plots

Let's start with a simple `ggplot` function. We are going to be adding layers to build up to more complex graphs, so we'll start by creating a base plotting object to build on.

```
BarPlot<-ggplot(aes(x=s, fill=Native), data=SelData)
```

5.6 aes

Recall from the *Basic Visualizations* Chapter, the use of aesthetic function `aes()`. This defines the data that we want to use for our `ggplot` graph. We will see how we do this by adding layers to our plot, similar to the way old-timey cartoons were made by layering multiple clear pages of cellophane with characters painted on them. The `aes` function inside of the `ggplot` function defines that data that will be shared among all of the layers. In addition, we can have separate `aes` functions inside different `geom_` layers that define and restrict the plotting data to that specific layer.

5.6. DISTRIBUTION PLOTS

Let's look at the ggplot object so far:

```
print(BarPlot)
```

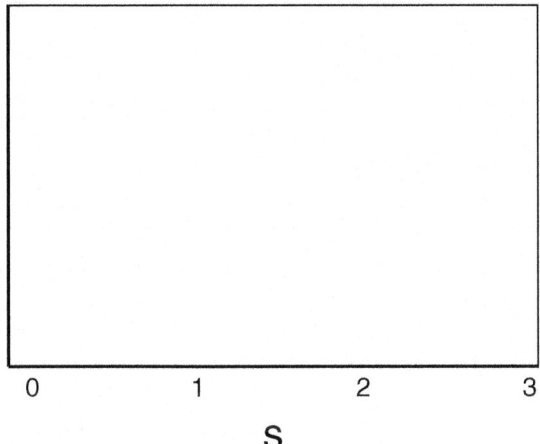

No data!

Question: Why are there no data plotted?

Answer: We didn't define a geom_ for the data yet.

5.6 Layers

So far, we have only loaded in the data for plotting. We have to specify which geom(s) we want. We'll start with a bar plot:

```
BarPlot<- BarPlot + geom_bar()
BarPlot
```

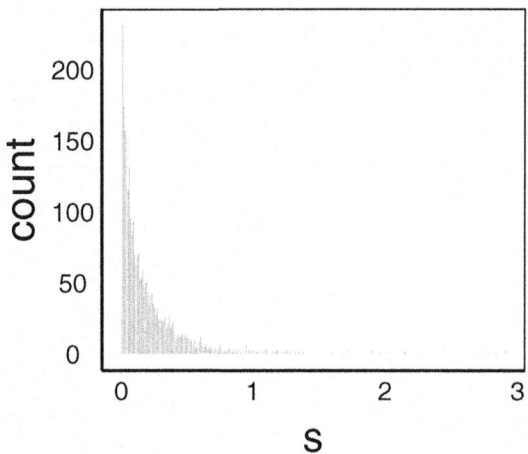

Let's explore the components of our Bar Plot object.

summary(BarPlot)

```
data: Collector, Author, Year, Journal, Vol, Species,
   Native, N, Fitness.measure, Trait, s, s.SE, s.P, B,
   B.SE, B.P, Rpoint, Rgroup [2766x18]
mapping:  x = ~s, fill = ~Native
faceting: <ggproto object: Class FacetNull, Facet, gg>
    compute_layout: function
    draw_back: function
    draw_front: function
    draw_labels: function
    draw_panels: function
    finish_data: function
    init_scales: function
    map_data: function
    params: list
    setup_data: function
    setup_params: function
    shrink: TRUE
    train_scales: function
    vars: function
```

5.6. DISTRIBUTION PLOTS

```
      super:  <ggproto object: Class FacetNull, Facet, gg>
-----------------------------------
geom_bar: just = 0.5, width = NULL, na.rm = FALSE, orientation = NA
stat_count: width = NULL, na.rm = FALSE, orientation = NA
position_stack
```

It's worth taking some time to understand the structure of this `ggplot` object how it relates to what gets plotted to the output.

1. **Data** shows which data are available for plotting. These are just the column names of the `data.frame` object we input with the `data=` parameter in the `ggplot()` function.
2. **Mapping** shows the variables from the `aes()` function, the scaling of the x-axis and the variable for the `fill=` colours.
3. **Faceting** contains information for multiple plots. We'll explore this more later, but a single graph has a `facet_null()` function.
4. **Dashed line** separates the `ggplot()` function from the other functions linked with the plus (+) in our plotting function. In our graph, there was + `geom_bar()`, which we can see below the dashed line.
5. **geom_bar** shows the (default) parameters used in our function
6. **stat_count** shows our `stat` function. This was created by default with our + `geom_bar()` function; it determines how the data are transformed to geometric shapes for plotting (e.g., points, lines, or bars).

The output also shows some of the functions and parameters used to generate the graph. At the bottom we see parameters for `geom_bar` and `stat_count`. Note that there are more parameters listed than what we explicitly put into the `ggplot()` function. These extra parameters are the **default parameters** for the function.

5.6 `geom_` and `stat_`

If `geoms` is the *geometry* of the shapes in the plot, then `stats` is the *statistic* or mathematical functions that create the geoms. In the above case, the vertical bars in `geom_bar` are created by counting the number of observations in each bin. The `stat_count` function is responsible for this calculation, and it is called by default when we use the `geom_bar` function. Specifically, `stat_count` *counts* the number of observations in each histogram bin.

More generally, we can change the geometry of the plotted shapes with `geom_<NAME>`, and we can define different functions for generating the geometric shapes with `stat_<NAME>`. To make things easier on us, there is a default `stat` for each `geom`. In most cases we can just focus on which geometry we want for our graph, and use the default `stat`.

For more information on the default parameters and `stat` of `geom_bar()` or any other geom, use the R help function.

```
?geom_bar
```

5.6 Bivariate geom

Let's explore a few more plotting options to get a better feel for our plotting parameters. Use `summary()` on each graph and compare it to the `summary()` output we examined earlier. Here we'll use the random normal values we generated above so that we can make a bivariate plot:

```
BivPlot<-ggplot(data=SelData, aes(x=s, y=Rpoint)) +
  geom_point()
```

First, take a quick look at the `summary()` of the plotting function and compare to the earlier graph (data not shown).

5.6. DISTRIBUTION PLOTS

```
summary(BivPlot)
```

Now plot the graph:

```
print(BivPlot)
```

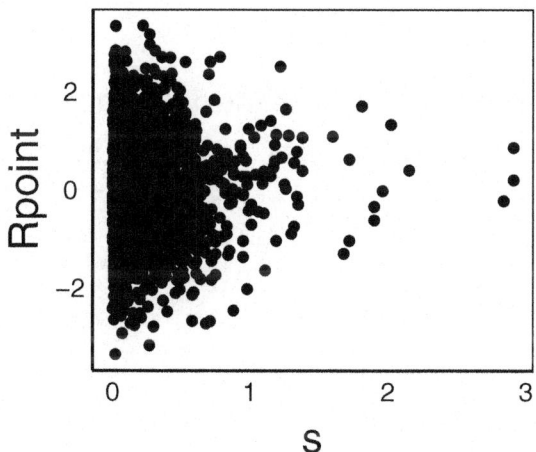

Notice how the points are all clustered to the left. This looks like a classic log-normal variable, so let's log-transform s (x-axis)

```
BivPlot<-ggplot(aes(x=log(s+1), y=Rpoint), data=SelData) +
  geom_point()
print(BivPlot)
```

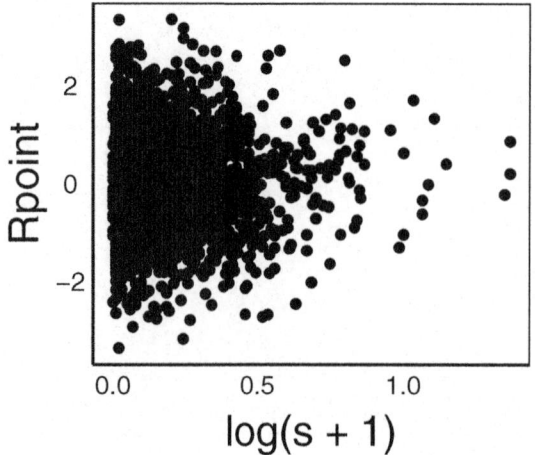

Once again, compare the summary with the untransformed x-axis.

> **Question**: What does the summary() show as the difference for a raw vs log-tranformed x-axis in the aes() function inside of ggplot()

5.6 geom_smooth

A really handy feature of ggplot is the geom_smooth function, which has several options for calculating and plotting a statistical model to the observations.

Here's a simple linear regression slope (lm = linear model):

```
BivPlot +
  geom_smooth(method="lm", colour="steelblue",
              formula = y~x, linewidth=2)
```

5.6. DISTRIBUTION PLOTS

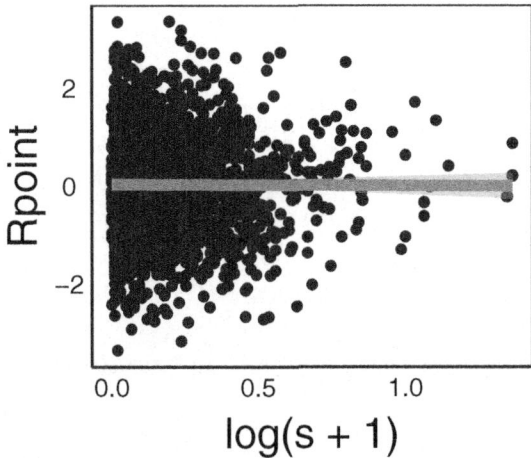

We can use a grouping variable to add separate regression lines for each group.

```
BivPlot +
  geom_smooth(aes(colour=Native),linewidth=2,
              method="lm", formula=y~x)
```

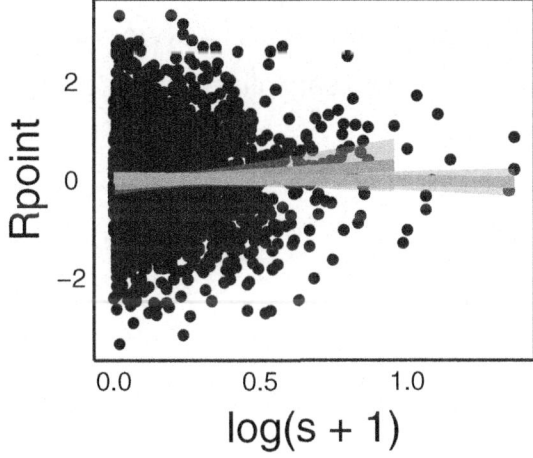

5.7 Full ggplot

Now that we have a better understanding of `ggplot()`, let's try to recreate the selection histograms in Figure 2 of Colautti & Lau (2015). This will involve three main steps:

1. Separate data for native vs. introduced species into two data sets for custom plotting.
2. Use a bootstrap model to estimate non-parametric mean and 95% confidence intervals for each group.
3. Plot all of the components on a single graph

One **technical note**. We are going to deviate from the published code slightly. The published figure uses *frequency* data, whereas we are going to use *density* data.

5.7 Separate data

Since this is a relatively simple resampling model, we'll use two separate vectors to store data for our plots and boostrap sampling: one for native and one for non-native.

```
NatSVals<-SelData$s[SelData$Native=="yes"]
IntSVals<-SelData$s[SelData$Native=="no"]
```

An alternative would be to set up a data frame and keep track of values as separate columns, with a different row for each iteration.

5.8 Bootstrap

The graph includes a bootstrap model to estimate the mean and variance for each group (`native` = "yes" vs "no"). A bootstrap is just a computational approach to generating confidence limits on the sample. It makes no assumptions about the underlying distribution of the population from which the sample is drawn, so it is a robust method for non-parametric data. In our case, we will randomly sample from each vector, and calculate the mean. We repeat this many times to get a range of values from which we can estimate the confidence interval.

The example below is not the most efficient approach but it is a good opportunity to practice our `for(){}` loops from the *Flow Control* section in the *R Fundamentals* Chapter.

5.8 Data Setup

First we define the number of iterations and set up two objects to hold the data from our bootstrap iterations.

```
IterN<-100 # Number of iterations
NatSims<-{} # Dummy objects to hold output
IntSims<-{}
```

5.8 `for` loop

We will use the `for` loop to resample the data, calcuate the sample mean, and repeat N times. This involves just three key steps.

1. Sample, with replacement.
2. Calculate the average.

3. Save the average in a vector: `NatSims` for native species or `IntSims` for non-native species.

```
for (i in 1:IterN){
  NatSims[i]<-mean(sample(
    NatSVals, length(NatSVals), replace=T))
  IntSims[i]<-mean(sample(
    IntSVals, length(IntSVals), replace=T))
}
```

Note in the above code we use 'nested' functions. The `sample()` function is nested inside the the `mean()` function, which is faster than using nested loop.

Also note that we can include both datasets (native + non-native) in the same `for` loop.

5.8 95% CI

Non-parametric **Confidence Intervals** (CI) are calculated directly from the bootstrap output. Let's try finding our 95% CI range, which goes from the lower 2.5% to the upper 97.5% of values.

First, sort the datea from low to high

```
NatSims<-sort(NatSims)
IntSims<-sort(IntSims)
```

Each of the output vectors contains a number of values equal to our `Iter` variable, as defined earlier in our code. Now we identify the lower 2.5% and upper 97.5% values in each vector. For example, with 1000 iterations our 2.5% would be the 25th value in the sorted vector and the upper 97.5% would be the 975th value in the sorted vector.

We use this number to index the vector with square brackets. We make sure to round to a whole number since we can't have a fractional cell position.

```
CIs<-c(NatSims[round(IterN*0.025,0)], # Lower 2.5%
       NatSims[round(IterN*0.975,0)], # Upper 97.5%
       IntSims[round(IterN*0.025,0)], # Lower 2.5%
       IntSims[round(IterN*0.975,0)]) # Upper 97.5%
```

The output (CIs) as is therefore a vector of four elements.

```
print(CIs)
```

```
[1] 0.1828284 0.2011856 0.2135459 0.3072541
```

Note: your numbers should be similar but won't be exact because you won't have the exact same random sample when you run your code.

> **Question**: What line of code could we add above to ensure that these numbers were exactly the same for everyone who ran this code?

Answer: If you aren't sure, then it's a good time to review the *R Fundamentals* Chapter.

5.9 Plot data

We'll set up a new data frame object for plotting data, to make it easier to write our plotting functions.

```
HistData<-data.frame(s=SelData$s,Native=SelData$Native)
```

and set up a minimal ggplot code:

```
p <- ggplot()
```

Now we can add layers to the plot. We'll print out each layer as we go, so that we can see what each layer adds to the overall graph. The coding is a bit complex here, so don't worry if it's hard to follow everything. The key thing to understand is how the different geoms contribute to the final plot.

```
p <- p + geom_density(aes(x=s),
                      data=HistData[HistData$Native=="yes",],
                      colour="#1fcebd66", size=2)
```

```
Warning: Using `size` aesthetic for lines was deprecated in ggplot2
3.4.0.
i Please use `linewidth` instead.
```

```
print(p)
```

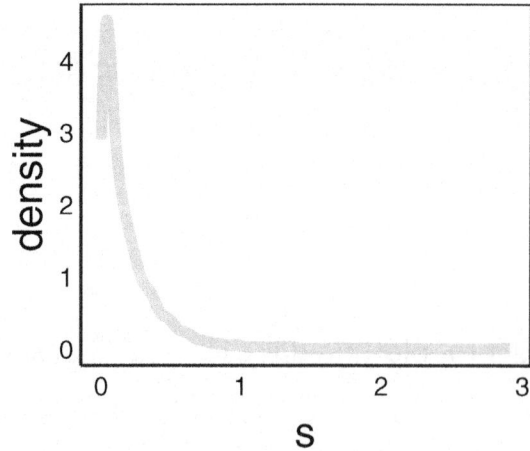

5.9. PLOT DATA

Here, we've added a geom_density geometry, which gives a smoothed line, like we saw in the *Quick Visualizations* Chapter. This works well for large datasets with many bins. For example, compare this graph with the geom_box() graph that we did on this data earlier in this chapter.

Note that the y-axis goes above 1 because the total probability is the area under the curve, which must equal one. Since there are many values that are less than one, the density must have values larger than 1 in order for the area to equal 1.

Next, we can add a similar graph for the non-native species, with a contrasting colour.

```
p <- p + geom_density(aes(x=s),
                data=HistData[HistData$Native=="no",],
             colour="#f5375166", size=2)
print(p)
```

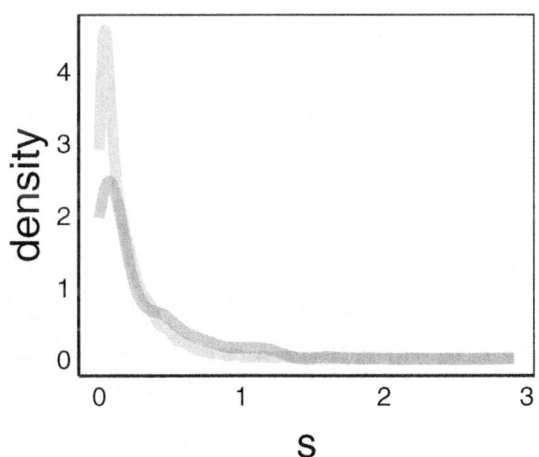

Alternatively, we could use a single geom with group= and colour=. However, using separate geom layers makes it easier to specify colours. This is yet another example of how different programming approaches can yield the same output, and one approach is not necessarily superior to the other.

Next, we add our CI data. CI is a range of values for each group, which we can represent as a rectangle.

```
p <- p + geom_rect(aes(xmin=CIs[1],xmax=CIs[2],ymin=0,ymax=0.25),
                   fill="#1fcebd88")
print(p) # native species 95% CI bar
```

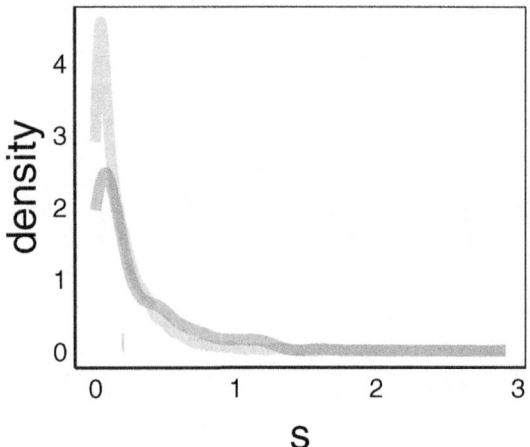

With geom_rect we define 4 points of the rectangle. The x-axes coordinates are the CI values from our bootstrap algorithm, and the y-axes values are arbitrary numbers that determine the height of the rectangle.

Now try the same thing for the non-native species.

```
p <- p + geom_rect(aes(xmin=CIs[3],xmax=CIs[4],ymin=0,ymax=0.25),
                   fill="#f5375188")
print(p) # introduced species 95% CI bar
```

5.9. PLOT DATA

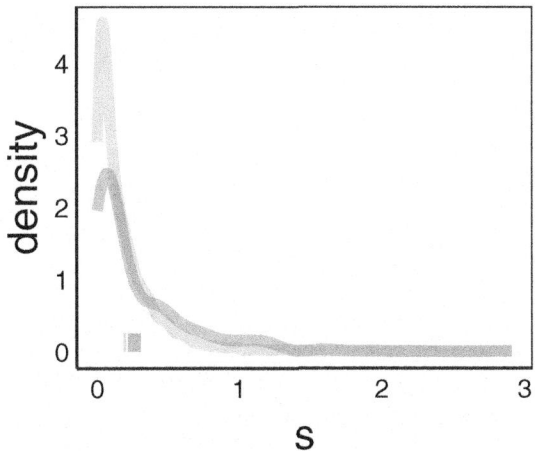

Now we add the bootstrap mean for each group. The bootstrap mean is just the mean of the bootstrap iterations – that is, a mean of means.

```
p <- p + geom_line(aes(x=mean(NatSims),y=c(0,0.25)),
                colour="#1d76bf",size=1)
p <- p + geom_line(aes(x=mean(IntSims),y=c(0,0.25)),
                colour="#f53751",size=1)
print(p)
```

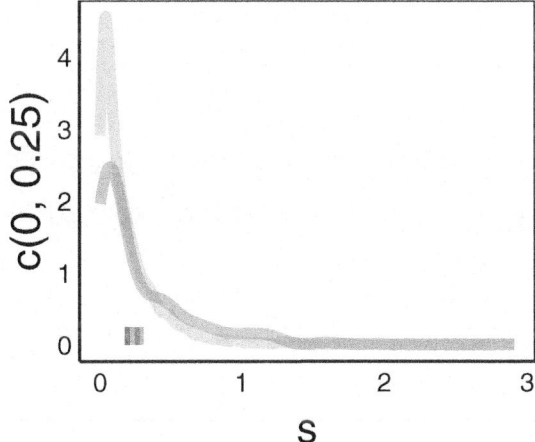

Finally, we tweak the axis titles and zoom in along the x-axis to make it easier to see the differences between the two distributions. Note how the y-axis label changed when we added geom_line.

```
p <- p + ylab("Density") +
   scale_x_continuous(limits = c(0, 1.5))
print(p) # labels added, truncated x-axis
```

```
Warning: Removed 12 rows containing non-finite values
(`stat_density()`).
```

```
Warning: Removed 1 rows containing non-finite values
(`stat_density()`).
```

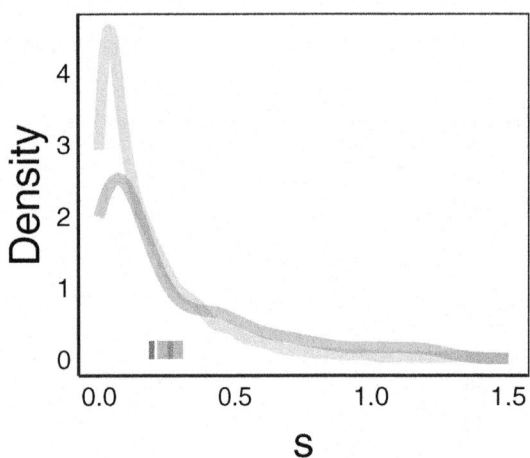

Note the warnings about missing data, which seems to be due to the fact that we zoomed in on the x-axis. Despite the warnings, everything looks okay if we compare this graph to the previous one.

There we have it! We succesffuly created a complex plot overlapping layers by breaking down ggplot() into individual components.

Chapter 6

Multi-plot Graphs

6.1 Overview

In the *Quick Visualizations* Chapter, we explored how to to make multiple graphs for different groups using the `facet()` function. Let's review briefly and then look at more advanced multi-graph options.

6.2 Setup

First, we'll run the usual plotting code.

```
library(ggplot2)
source("http://bit.ly/theme_pub")
theme_set(theme_pub())
```

Continuing from the previous chapter, we'll work with the selection data from Colautti & Lau (2015). We'll also change the header names, remove missing values, replace s with $|s|$ and add a random variable `Rpoint`, as we did in the previous chapter.

```
SelData<-read.csv(
  "https://colauttilab.github.io/RCrashCourse/Selection_Data.csv")

names(SelData)<-c("Collector", "Author", "Year", "Journal",
                  "Vol", "Species", "Native", "N",
                  "Fitness.measure", "Trait", "s",
                  "s.SE", "s.P", "B", "B.SE", "B.P")
SelData<-SelData[!is.na(SelData$s),]
SelData$s<-abs(SelData$s)
SelData$Rpoint<-rnorm(nrow(SelData))
```

6.3 facets

There are three main facet functions, each with different options.

1. `facet_null` makes a single graph, this is the default for ggplot(), as we saw earlier
2. `facet_grid` lets us define a grid and set the vertical and horizontal variables
3. `facet_wrap` is a convenient option if only have one categorical variable but many categories

Remember: one little tricky part of facets with `ggplot` is that we can either use the tilde notation (. ~ .) with the `facet_grid` function, or else we must define the variable for faceting with the `vars()` function. The `vars()` function indicate which categorical variables from the original data set should be used to subset the graphs.

Returning to the `BivPlot` example above:

6.3. FACETS

```
BivPlot<-ggplot(data=SelData, aes(x=log(s+1), y=Rpoint)) +
  geom_point()
BivPlot + facet_grid(Native ~ Collector)
```

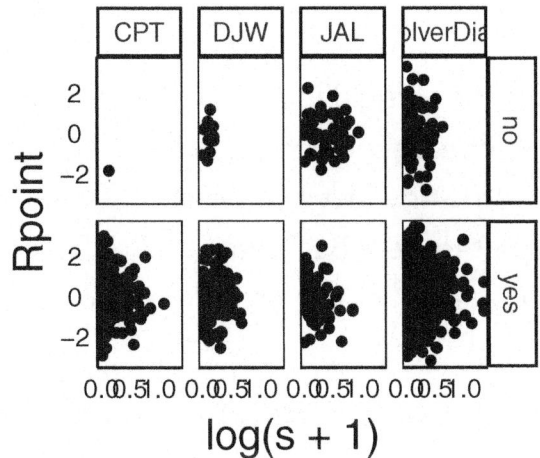

```
BivPlot<-ggplot(data=SelData, aes(x=log(s+1), y=Rpoint)) +
  geom_point()
BivPlot + facet_wrap(vars(Year))
```

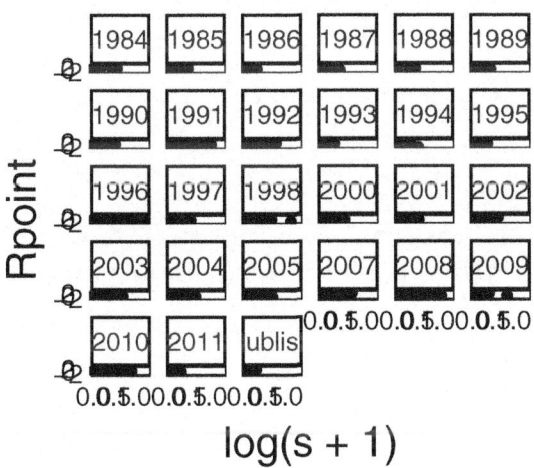

Note that this large, multi-panel graph does not reproduce well in this textbook, but may look better if plotted to a large window on your computer, or output to an external file with the pdf() or svg() functions, as discussed in the *Basic Customizations* Chapter.

6.4 `gridExtra` package

Facets produce graphs that all have the same dimension and the same x- and y-axes. We might call these 'homogeneous' plots because they use a homogeneous format. For some advanced publications and reports, we might want to include 'heterogeneous' plots with different axes and different sizes. The `gridExtra` package provides options for this.

Remember to install with `install.packages("gridExtra")` before you try to load the library for the first time.

```
library(gridExtra)
```

```
Attaching package: 'gridExtra'

The following object is masked from 'package:dplyr':

    combine
```

The `grid.arrange()` funcrion from the `gridExtra` package allows for more complex multi-panel figures.

6.4 `grid.arrange()`

Use this to combine heterogeneous ggplot objects into a single multi-panel plot.

6.4. GRIDEXTRA PACKAGE

Note that this will print graphs down rows, then across columns, from top left to bottom right. You can use nrow and ncol to control the layout in a grid format.

```
HistPlot<-ggplot(aes(x=s,colour=Native), data=SelData) +
  geom_density()
BarPlot<-ggplot(aes(x=s,fill=Native), data=SelData) +
  geom_histogram(binwidth=1/10)
grid.arrange(HistPlot,BivPlot,BarPlot,ncol=1)
```

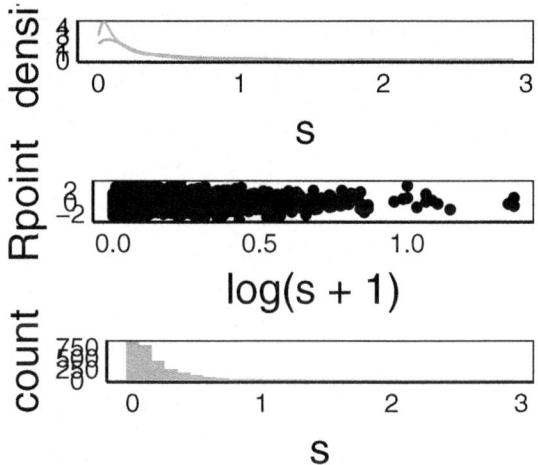

```
grid.arrange(HistPlot,BivPlot,BarPlot,nrow=2)
```

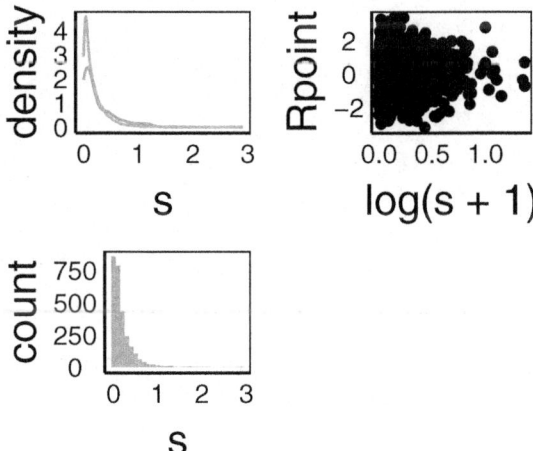

You might get some warnings based on missing values or wrong `binwidth` options. You will also see some weird things with different text sizes in the graphs. Normally, you would want to fix these for a final published figure but here we are just focused on showing what is possible with the layouts.

We can see that `grid.arrange()` allows us to combine multiple graphs with different axes, data, and `geom_` geometries. However, the layout of the graphs all have the same dimension. What if we want to combine plots of different size? For example, maybe we want to have one graph that is narowwer but wider than the other two. Or maybe we would like to inset a smaller graph inside of a larger one. The `grid` package can handle this.

6.5 `grid` package

We can make more advanced multi-panel graphs using the `grid` package. This is part of the base installation of R so you don't need to use `install.packages()` this time.

```
library(grid)
```

First, we set up a new plotting area with `grid.newpage()`.

```
grid.newpage() # Open a new page on grid device
```

You won't see anything plotted yet. To insert a new graph on top (or inside) the current graph, we use `pushViewport` to set up an imaginary plotting grid. In this case, imagine breaking up the plotting space into 3 rows by 2 columns.

6.5. GRID PACKAGE

```
pushViewport(viewport(layout = grid.layout(3, 2)))
```

Again, there is nothing being plotted yet, we have only set up the plotting area. Next, we print each plotting object into the grid(s) space we would like it to go.

Add the first figure in row 3 and across columns 1:2

```
print(HistPlot, vp = viewport(layout.pos.row = 3,
                              layout.pos.col = 1:2))
```

Add the next figure across rows 1 and 2 of column 1

```
print(BivPlot, vp = viewport(layout.pos.row = 1:2,
                             layout.pos.col = 1))
```

Add the final figure across rows 1 and 2 of column 2

```
print(BarPlot, vp = viewport(layout.pos.row = 1:2,
                             layout.pos.col = 2))
```

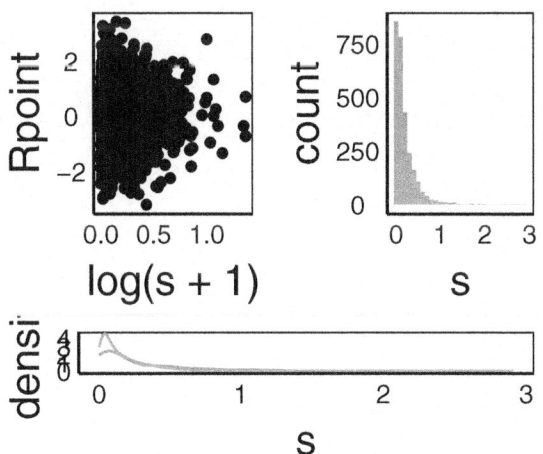

6.5 Inset

We can also use `pushViewport` to set up a grid for plotting on top of an existing graph or image. This can be used to generate a figure with an inset.

First generate the 'background' plot. Note that you could alternatively load an image here to place in the background.

```
HistPlot
```

Next, overlay an invisible grid layout, with the number of cells that can be used to determine size and location of the inset graph. In this case, we'll set up a 4-by-4 grid and then plot in the top, right corner.

```
pushViewport(viewport(layout = grid.layout(4, 4)))
```

Finally, add the graph. In this case we want it only in the top two rows and the right-most two columns – i.e. the top-right corner.

```
print(BivPlot, vp = viewport(layout.pos.row = 1:2, layout.pos.col = 3:4))
```

The final product:

```
HistPlot
pushViewport(viewport(layout = grid.layout(4, 4)))
print(BivPlot, vp = viewport(layout.pos.row = 1:2, layout.pos.col = 3:4))
```

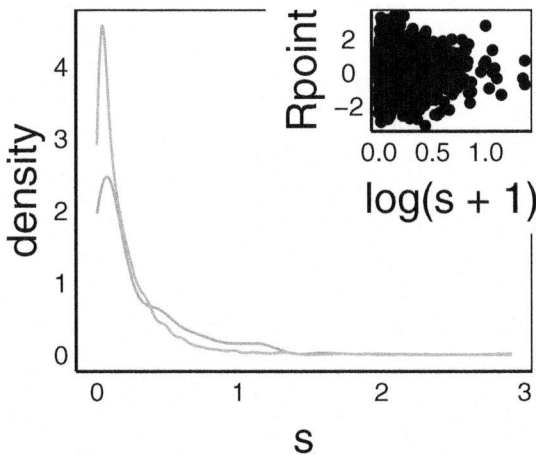

6.6 Further Reading

The 2009 book *ggplot2: Elegant Graphics for Data Analysis* by Hadley Wickham is the definitive guide to all things ggplot.

A physical copy is published by Springer: *http://link.springer.com/book/10.1007%2F978-0-387-98141-3*

And there is a free ebook version: *https://ggplot2-book.org/*

Chapter 7

Regular Expressions

7.1 Overview

Regular Expressions, also known as **regex** and **regexp** are special text-based functions that act as run complex find-and-replace functions. I didn't learn regular expressions until I was a postdoc working at Duke University, but I wish I had learned about them much earlier! This remains one of the most useful programming tools I have ever used. It is absolutely essential for working with any kind of large text files or large data sets. I'll explain.

A lot of programming tools in biology use input text files that require very specific formatting (e.g. .txt, .csv, .fasta, .nex). Sometimes, you might need to reorganize or recode data in a large text file or in many separate text files. This can be a big time sink, it can introduce errors, and it's not reproducible if you do it manually. But regular expressions can automate the process.

Here's one example. As a PhD student I co-founded a project called the **Global Garlic Mustard Field Survey (GGMFS)** with collaborator Dr. Oliver Bossdorf at the University of Tübingen – yes the same Dr. Bossdorf mentioned in the *Quick Visualizations* Chapter.

We were fortunate to have over 100 collaborators across Europe and North America who helped to collect samples for the project. Details of the project were published in the Journal **Neobiota**: *https://neobiota.pensoft.net/article/1270/* but one BIG problem is the way that each of these 100+ collaborators entered their data online. For example, latitudes and longitudes were entered in a variety of different formats. Regular expressions allowed me to write a small program to automatically convert all of these different formats to a common, decimal format that we could use for the analysis. This saved a huge amount of time and prevented errors that could have been introduced if we tried to edit these values by hand.

Often when you work with large datasets, you will need to automate some of your error correction, and regular expressions can be a big help here. For example, imagine a simple online survey that includes a place for people to simply type "yes" or "no" in response to a question. This should be coded as a binary variable (1 or 0) for analysis, but you might find a variety of inputs such as: "YES", "Y", "yes", and "Yes". These all mean the same thing, yet if you try to analyze the raw output, R will treat these as different categories. Here again, regular expressions can be used to quickly change all the different examples to a common "Y" or to a Boolean variable TRUE.

One final example, is pattern matching, which is common for the analysis of DNA, proteins or other large strings of data. You may want to find a particular sequence of data, possibly with a few variable sites: e.g. TCTA or TCAA or TCGA. This is another area where regular expressions can help.

7.1 Universal Syntax

Regular expressions are a universal language that extends to many other programming languages, including **C/C#/C++**, **Python**, **Unix/Linux**, and **Perl**. We focus here on R but most of the syntax is mantained across programming languages.

7.1 Steep Learning Curve

WARNING! There is a very steep learning curve here, and the only way to really learn this is to drown yourself in examples. There are lots of exercises you can do for practice online. You should also try to apply these whenever you can, just like you should with all of your other R skills.

7.2 Functions

There are four main functions that use regular expressions in R.

`grep()` and `grepl()` are equivalent to 'find' in your favorite word processor. They have the general form:

```
gsub("find", in.this.object)
```

`grep()` outputs a vector with all of the address locations (i.e. numbers) that match. Thus the *output length* is equal to the *number of matches*.

`grepl()` outputs a vector of TRUE (match) and FALSE (no match). Thus, the *output length* is equal to the *length of the input object*.

`sub()` and `gsub()` are equivalent to 'find and replace'. They have the general form:

```
grep("find", "replace", in.this.object)
```

`sub()` replaces only the first match, whereas `gsub()` replaces all of the matches.

Some specific examples are provided below to help you understand these similarities and differences. As always, you should take the time to try these out and make sure you get the same input. If you don't, then it's a good learning opportunity to find out what you did differently!

There are two other more advanced functions in R. These aren't covered in this tutorial, but may be of use once you are more comfortable with the above functions.

`regexpr()` provides more detailed information about the first match.

`gregexpr()` provides more detailed results about all matches.

> See `?regexpr` and `?gregexpr` for more info

7.2 Examples

Some examples can help to understand the differences among the four main functions. Let's start with a simple data frame of species names.

```
Species<-c("petiolata", "verticillatus", "salicaria", "minor")
print(Species)
```

```
[1] "petiolata"     "verticillatus" "salicaria"
[4] "minor"
```

7.2 grep()

This returns cell addresses matching the query string.

grep("a",Species)

[1] 1 2 3

Note the vector length compared to the input vector. Instead of the cell number, we can get R to return the specific values in each matching cell with the value=T parameter

grep("a",Species, value=T)

[1] "petiolata"   "verticillatus" "salicaria"

7.2 grepl()

This returns a vector of TRUE (match) and FALSE (no match). Compare this output with the same parameters in the grep() function.

grepl("a",Species)

[1]   TRUE   TRUE   TRUE FALSE

7.2 sub()

This replaces the first match (in each cell)

```
sub("l","L",Species)
```

```
[1] "petioLata"     "verticiLlatus" "saLicaria"
[4] "minor"
```

7.2 gsub()

This replaces all matches (in each cell). Compare this output to sub().

```
gsub("l","L",Species)
```

```
[1] "petioLata"     "verticiLLatus" "saLicaria"
[4] "minor"
```

Question: Did you see the difference?

Hint: Look at "Verticillatus".

7.3 Wildcards

7.3 \ escape character

The backslash is a special character. It's called the 'escape' character because it is used to *escape* from the literal interpretation of the next character to the right. For example, \. applies the *escape* to the period character. The specific meaning depends on the context, which is much easier to understand by examples, as shown below.

7.3 \\ in R

In the introduction, we discussed the universality of **regular expressions** in the sense that a similar syntax is used by many different programming langagues. But now here is one exception. In R, the double-escape is usually needed, whereas other programming languages typically use just one. The reason is a bit meta – it's because we are running regular expressions within R object. So the first \ is used to escape special characters in R, applying it to the second \, which is itself the special character that needs to be escaped to pass through the function. The second slash is followed by the 'escaped' character. Some examples are provided below.

If that isn't clear. Just remember that you need two backslashes when writing regular experssions in R, but just one backslash for most other languages.

7.3 \\w

Instead of finding the letter w, the \\w is a **wildcard** character that represents any letter or digit. It also includes underscore _ for some reason.

```
sub("w","X","...which 1-100 words get replaced?")
```

```
[1] "...Xhich 1-100 words get replaced?"
```

```
gsub("w","X","...which 1-100 words get replaced?")
```

```
[1] "...Xhich 1-100 Xords get replaced?"
```

```
sub("\\w","X","...which 1-100 words get replaced?")
```

```
[1] "...Xhich 1-100 words get replaced?"
```

```
gsub("\\w","X","...which 1-100 words get replaced?")
```

```
[1] "...XXXXX X-XXX XXXXX XXX XXXXXXXX?"
```

Again, note the differences between the `sub()` and `gsum()` functions. We'll stick to `gsub()` for the remainder of the examples in this chapter, but you should also run `sub()` yourself. Each time, take a moment to try to predict how the output will differ before running it. This will help you develop an understanding of regular expressions much more quickly.

7.3 \\W

The capital W is the inverse of \\w find a character that is NOT a letter or number.

```
gsub("\\W","X","...which 1-100  words get replaced?")
```

```
[1] "XXXwhichX1X100XXwordsXgetXreplacedX"
```

7.3 \\s

This represents a space

```
gsub("\\s","X","...which 1-100  words get replaced?")
```

```
[1] "...whichX1-100XXwordsXgetXreplaced?"
```

7.3 \\t

This is a tab character. A lot of data files stored as text are tab-delimited (.tsv) as well as comma-delimited (.csv)

gsub("\\t","X","...which 1-100 \t words get replaced?")

[1] "...which 1-100 X words get replaced?"

Remember that \t is a tab character.

cat("A\t\t\tB C")

A   B C

7.3 \\d

d for *digits*. This is the wild card for numeric characters.

gsub("\\d","X","...which 1-100 words get replaced?")

[1] "...which X-XXX words get replaced?"

7.3 \\D

Non-digit characters

gsub("\\D","X","...which 1-100 words get replaced?")

[1] "XXXXXXXXX1X100XXXXXXXXXXXXXXXXXXXXXX"

7.4 New Lines

There are two special characters that indicate new lines in a text file.

7.4 \\r

This is the 'carriage return' special character

7.4 \\n

This is the 'newline' special character

7.4 Big Problem

One or both of these may be generated when you press the 'enter' key while writing a text file. The difference depends on which operating system you are using. These also add a source of headache and confusion when working with text files because:

1. **Unix and MacOS** text files use lines that end with \n only
2. **Windows and DOS** text files use lines end with \r\n

> **Question**: Do you know how this difference originated?

Answer: The reason goes back to the early days of programming, when programmers were moving from mechanical typewriters to computer programs. Mechanical typewriters are hard to find these days, but they would hold a piece of paper in place on a cylinder called a *carriage*. The

The \n stands for 'new line', and the \r stands for *return*. When you reach the end of a line of text on a typewriter, you would typically *return* the carriage back to the starting position, and then move to the next line, thus the \r\n. The Unix operating system decided that the \r wasn't needed, whereas the DOS operating system decided to include it.

This difference can cause problems when moving text files across operating systems. Programs like *FileZilla* will automatically translate these end-of-line characters when moving across systems.

7.5 Special characters

In addition to special characters that use the escape \\, there are a number of other special characters that don't use the escape, but have a special meaning.

Note that if you want to search for the characters below you would have to use the escape character. E.g., use \\. to search for the period character (.).

7.5 . (any character)

The period is a wild card that means 'anything'. This includes all of the \\w characters but also other characters like puncutation marks.

```
gsub(".","X","...which 1-100  words get replaced?")
```

[1] "XXXXXXXXXXXXXXXXXXXXXXXXXXXXXXXXX"

So how to search for a period .? As noted above, we have to use the escape character

```
gsub("\\.","X","...which 1-100  words get replaced?")
```

```
[1] "XXXwhich 1-100  words get replaced?"
```

7.5 | (or)

This is sometimes called the **pipe** character, and it simply means 'or'. For example, we can search for w or e.

```
gsub("w|e","X","...which 1-100  words get replaced?")
```

```
[1] "...Xhich 1-100  Xords gXt rXplacXd?"
```

7.5 *, ?, +, {} (special searches)

These special characters refer to details about the kind of search that we are trying to conduct. Look at these examples carefully, and remember that sub replaces the first match while gsub replaces all of the matches.

```
sub("\\w","X","...which 1-100 words get replaced?")
```

```
[1] "...Xhich 1-100 words get replaced?"
```

```
gsub("\\w","X","...which 1-100 words get replaced?")
```

```
[1] "...XXXXX X-XXX XXXXX XXX XXXXXXXX?"
```

Now let's apply some of these special characters to see how they work.

7.5 + (find *one or more* matches)

Finds 'one or more' matches (i.e. at least one match)

```
sub("\\w+","X","...which 1-100 words get replaced?")
```

```
[1] "...X 1-100 words get replaced?"
```

```
gsub("\\w+","X","...which 1-100 words get replaced?")
```

```
[1] "...X X-X X X X?"
```

Compare this match to the one above. Notice how we have replaced groups of letters instead of single letters. The algorithm works like this:

1. Start at the left and move to the right, one character at a time
2. Check if the character is a letter or number (\\w).
3. If NO, move to the next character
4. If YES, check the next character. If it is also a \\w then go to the next character. Repeat until the next character is not \\w, and replace the entire string of characters.

When run in the `sub()` function, the algorithm does the above and then stops. When run with the `gsub()` function, it continues to the next character, and then starts over.

7.5 * (*greedy* matches)

This is a *greedy* search matches *0 or more* in a row. Again, this is easier to understand by exploring examples.

```
sub("\\w*","X","...which 1-100 words get replaced?")
```

```
[1] "X...which 1-100 words get replaced?"
```

```
gsub("\\w*","X","...which 1-100 words get replaced?")
```

```
[1] "X.X.X.X X-X X X X?X"
```

In the `sub()` function, it detects a period (.) as the first character, indicating no match. It replaces the 'null' or 0 match at the beginning, which has the effect of adding a character. In the `gsub()` function it repeats this again before each period (.). It then continues until it finds the letter w. Then it finds a group of \\w matches, replacing all of them with a single X. Then a space, which is skipped, then a -, which is another null match, prompting another insert.

7.5 ? (*restrained* match)

This is the *restrained* search, which matches *zero or one* time.

```
sub("\\w?","X","...which 1-100 words get replaced?")
```

```
[1] "X...which 1-100 words get replaced?"
```

```
gsub("\\w?","X","...which 1-100 words get replaced?")
```

```
[1] "X.X.X.XXXXX X-XXX XXXXX XXX XXXXXXXX?X"
```

Compare this to the * above. The ? character behaves in a similar way, except it is constrained in the sense that each each letter is replaced individually, instead of replacing entire words.

7.5 +? (*lazy, restrained)

This is the *lazy* version of +

```
sub("\\w+?","X","...which 1-100 words get replaced?")
```

```
[1] "...Xhich 1-100 words get replaced?"
```

```
gsub("\\w+?","X","...which 1-100 words get replaced?")
```

```
[1] "...XXXXX X-XXX XXXXX XXX XXXXXXXX?"
```

Note the difference in sub(), which replaces on the the first letter here but the whole word when + is used alone in the earlier example. In the gsub() example we end up replacing every letter instead of whole words. Remember, sub() runs the algorithm once and then stops, while gsub() cycles through the algorithm until it reaches the end of the line.

7.5 *? (*lazy, greedy)

Similarly, we can combine these characters for the 'lazy' version of *

```
sub("\\w*?","X","...which 1-100 words get replaced?")
```

```
[1] "X...which 1-100 words get replaced?"
```

```
gsub("\\w*?","X","...which 1-100 words get replaced?")
```

```
[1] "X.X.X.XwXhXiXcXhX X1X-X1X0X0X XwXoXrXdXsX XgXeXtX XrXeXpXlXaXcXeXdX?X"
```

Try using +*.

Question: Why do you get an error message?

Answer: The * and ? find the same characters, but have competing replacement rules (greedy or restrained).

7.5 {} (range)

Curly brackets are used to specify a number of matches, expanding on the options even further.

7.5 {n,m}

Find between n to m matches

```
gsub("\\w{3,4}","X","...which 1-100 words get replaced?")
```

```
[1] "...Xh 1-X Xs X XX?"
```

7.5 {n}

Find exactly n matches

```
gsub("\\w{3}","X","...which 1-100 words get replaced?")
```

```
[1] "...Xch 1-X Xds X XXed?"
```

7.5 {n,}

Find n or more matches

```
gsub("\\w{4,}","X","...which 1-100 words get replaced?")
```

[1] "...X 1-100 X get X?"

7.5 {}?

As above, we can use ? for the 'lazy' versions of these searches

```
gsub("\\w{4,}?","X","...which 1-100 words get replaced?")
```

[1] "...Xh 1-100 Xs get XX?"

7.6 [ ] (set)

Square brackets allow us to define a *set*, which is a group of characters from which we want to match *any*. Within a set, we can use the dash - to specify a range of numbers or letters.

```
gsub("[aceihw-z]","X","...which 1-100 words get replaced?")
```

[1] "...XXXXX 1-100 Xords gXt rXplXXXd?"

In the above example, we search for 1 of any of the listed letters: a, c, e, i h, w, x, y, z. Note that x and y are included in the w-z statement.

> **Question**: What if we want to find 1 or more of these characters in a row to replace with X?

```
gsub("[aceihw-z]+","X","...which 1-100 words get replaced?")
```

```
[1] "...X 1-100 Xords gXt rXplXd?"
```

7.7 ^ (start or negate) and $ (end)

Use these characters to specify searches at the start ^ or end $ of the input string.

7.7 ^ (start)

How do we find which species start with the letter *a*? Use the start character:

```
grep("^a",Species)
```

```
integer(0)
```

7.7 ^ (negate)

The same character (^) has a different meaning if used with a set [ ]. In those cases, it negates, or finds the opposite.

For example, find species containing any character other than *a*:

```
grep("[^a]",Species)
```

```
[1] 1 2 3 4
```

Replace every letter except *a* or *l*

```
gsub("[^al]","X",Species)
```

```
[1] "XXXXXlaXa"      "XXXXXXXllaXXX" "XalXXaXXa"
[4] "XXXXX"
```

7.7 $ (end)

Find species that end with *a*

```
grep("a$",Species)
```

```
[1] 1 3
```

7.8 ( ) (capture)

Regular parentheses are used to 'capture' text, which can then be specified in the replacement string using \\1. Or you can capture multiple pieces of text and reorganize them by using the corresponding number – \\1 for the first set of ( ), \\2 for the second set of ( ), etc. Some examples should help.

Replace each word with its first letter

```
gsub("(\\w)\\w+","\\1",
     "...which 1-100 words get replaced?")
```

```
[1] "...w 1-1 w g r?"
```

Pull out only the numbers and reverse their order

```
gsub(".*([0-9]+)-([0-9]+).*",
      "\\2-\\1","...which 1-100 words get replaced?")
```

[1] "100-1"

Reverse first two letters of each word

```
gsub("(\\w)(\\w)(\\w+)","\\2\\1\\3",
     "...which 1-100 words get replaced?")
```

[1] "...hwich 1-010 owrds egt erplaced?"

7.9 Scraping

Scraping is a method for collecting data from online sources. In R, we can use the functions `readLines` and `curl()`, both from the `curl` library, to copy data from websites. We can do this because websites with the `.html` or `.xml` extension are a special kind of text files.

For more advanced applications, we might want to use the `rvest` package, which is designed for `html` and `xml` files. However, we'll focus here on `curl()` because we can apply regular expressions to any text files to extract information of interest. Here's an example where we will scrape a record for the Green Fluorescent Protein (GFP) from the Protein Data Bank (PDB). Note that this is a file with the extension `.pdb` but this is a human-readable text file that can be opened in any text editor

First, we'll import the text into an R object.

```
library(curl)
```

Using libcurl 7.64.1 with Schannel

7.9. SCRAPING

You will have to use `install.packages("curl")` to download this package to your computer. You only need to do this once but you will have to use `library(curl)` whenever you want to use the functions, as explained in the *R Fundamentals* Chapter.

Now we can download a file to play with.

```
Prot<-readLines(curl("http://www.rcsb.org/pdb/files/1ema.pdb"))
```

Download this link to your computer and open with a text file to see what it looks like.

This hint is a simple trick to understand what kind of file(s) you are working with.

This is a tab-delimited file, which we could import as a data frame using `read.delim` but we'll keep it this way to practice our regular expressions.

The `Prot` object we have made is a simple vector of strings, with each cell corresponding to a different row of text:

```
length(Prot)
```

[1] 2363

```
grep("TITLE",Prot)
```

[1] 2

We can pull out the amino acid sequences, which are rows that start with the word 'ATOM'

```
AAseq<-Prot[grep("^ATOM",Prot)]
length(AAseq)
```

```
[1] 1717
```

```
AAseq[1]
```

```
[1] "ATOM      1  N   SER A   2      28.888   9.409  52.301  1.00 85.05           N "
```

Challenge: Try to apply what you have learned about regular expressions to isolate the 3-letter amino acid code.

There are several ways we could do this. Take the time to think about it and give it a try.

Here's one good option, since we know it's a tab-delimited file with the amino acid in the 4th column:

```
gsub("ATOM\\t\\w+\\t\\w+\\t(\\w+).*","\\1",AAseq[1])
```

```
[1] "ATOM      1  N   SER A   2      28.888   9.409  52.301  1.00 85.05           N "
```

That didn't work. Sometimes the 'tabs' are actually just multiple 'spaces'

```
AAchain<-gsub("ATOM\\s+\\w+\\s+\\w+\\s+(\\w+).*","\\1",AAseq)
AAchain[1:100]
```

```
 [1] "SER" "SER" "SER" "SER" "SER" "SER" "LYS" "LYS" "LYS"
[10] "LYS" "LYS" "LYS" "LYS" "LYS" "LYS" "GLY" "GLY" "GLY"
[19] "GLY" "GLU" "GLU" "GLU" "GLU" "GLU" "GLU" "GLU" "GLU"
[28] "GLU" "GLU" "GLU" "GLU" "GLU" "GLU" "GLU" "LEU" "LEU"
```

```
 [37] "LEU" "LEU" "LEU" "LEU" "LEU" "LEU" "PHE" "PHE" "PHE"
 [46] "PHE" "PHE" "PHE" "PHE" "PHE" "PHE" "PHE" "PHE" "THR"
 [55] "THR" "THR" "THR" "THR" "THR" "THR" "GLY" "GLY" "GLY"
 [64] "GLY" "VAL" "VAL" "VAL" "VAL" "VAL" "VAL" "VAL" "VAL"
 [73] "VAL" "VAL" "VAL" "VAL" "VAL" "VAL" "PRO" "PRO" "PRO"
 [82] "PRO" "PRO" "PRO" "PRO" "ILE" "ILE" "ILE" "ILE" "ILE"
 [91] "ILE" "ILE" "ILE" "LEU" "LEU" "LEU" "LEU" "LEU" "LEU"
[100] "LEU"
```

Now we have a handy vector of amino acids representing our protein.

7.10 Examples

Let's try practising with a couple of examples.

7.11 Transect Data

Regular expressions are also useful with data objects

Imagine you have a repeated measures design. 3 transects (A-C) and 3 positions along each transect (1-3). We can simulate this data by generating dandom numbers in a data frame.

```
Transect<-data.frame(Species=letters[1:20],
                     A1=rnorm(20), A2=rnorm(20), A3=rnorm(20),
                     B1=rnorm(20), B2=rnorm(20) ,B3=rnorm(20),
                     C1=rnorm(20), C2=rnorm(20), C3=rnorm(20))
head(Transect)
```

```
  Species         A1          A2         A3          B1
1       a -0.4198256 -0.13312498 -1.4109151  0.79695255
2       b  0.9787430  1.19501106 -1.2830537 -0.44903314
```

```
3          c -1.3156918  0.09378044  0.7336550  0.02955709
4          d  0.4451767 -0.33857022  1.2335349  0.73913449
5          e -0.1564192  1.97034831  0.8966604 -0.94997272
6          f  0.2982561  1.14726923  0.9084409 -0.40926605
          B2           B3           C1           C2           C3
1 -0.19772971  0.7598830  0.1652908  1.5544440 -1.1076870
2 -1.44271844  0.6350828  0.2532211 -1.8921375 -0.3547272
3  0.84646565  0.3568771 -1.5541517  1.7620628  1.7550873
4  0.40875611  0.7041051 -0.2428876  0.8681144 -1.7863549
5 -1.69313811 -1.4492467  0.6620788 -1.8923036 -1.9831389
6  0.05911306  0.7973128 -3.2050414 -0.3922813  2.1686814
```

Tip: the object `letters` in the code above contains lower-case letter, while LETTERS contains upper case.

7.11 Challenge

Use your knowledge f regular expressions with subsetting data outlined in the *R Fundamentals* Chapter to do the following:

1. Subset only the columns that have an "A" in their name
2. Subset the rows of data for species "d"

Take the time to do this on your own. It will take you a while and you might make a lot of mistakes along the way. That's all part of the learning process. The longer you struggle, the faster you will learn.

Now here is a more challenging example:

7.12 Genbank

Here is a line of code to import DNA from genbank. (The one line is broken up into three physical lines to make it easier to read)

7.12. GENBANK

```
Lythrum_18S<-scan(
  "https://colauttilab.github.io/RCrashCourse/sequence.gb",
  what="character",sep="\n")
```

This is the sequence of the 18S subunit from the ribosome gene of *Lythrum salicaria* (from Genbank)

```
print(Lythrum_18S)
```

```
 [1] "LOCUS       AF206955                1740 bp    DNA     linear   PLN 18-APR-2003"
 [2] "DEFINITION  Lythrum salicaria 18S ribosomal RNA gene, complete sequence."
 [3] "ACCESSION   AF206955"
 [4] "VERSION     AF206955.1"
 [5] "KEYWORDS    ."
 [6] "SOURCE      Lythrum salicaria"
 [7] "  ORGANISM  Lythrum salicaria"
 [8] "            Eukaryota; Viridiplantae; Streptophyta; Embryophyta; Tracheophyta;"
 [9] "            Spermatophyta; Magnoliopsida; eudicotyledons; Gunneridae;"
[10] "            Pentapetalae; rosids; malvids; Myrtales; Lythraceae; Lythrum."
[11] "REFERENCE   1  (bases 1 to 1740)"
[12] "  AUTHORS   Soltis,P.S., Soltis,D.E. and Chase,M.W."
[13] "  TITLE     Direct Submission"
[14] "  JOURNAL   Submitted (19-NOV-1999) School of Biological Sciences, Washington"
[15] "            State University, Pullman, WA 99164-4236, USA"
[16] "FEATURES             Location/Qualifiers"
[17] "     source          1..1740"
[18] "                     /organism=\"Lythrum salicaria\""
[19] "                     /mol_type=\"genomic DNA\""
[20] "                     /db_xref=\"taxon:13129\""
[21] "                     /note=\"Lythrum salicaria L.\""
[22] "     rRNA            1..1740"
[23] "                     /product=\"18S ribosomal RNA\""
[24] "ORIGIN      "
[25] "        1 gtcatatgct tgtctcaaag attaagccat gcatgtgtaa gtatgaacaa attcagactg"
[26] "       61 tgaaactgcg aatggctcat taaatcagtt atagtttgtt tgatggtatc tgctactcgg"
[27] "      121 ataaccgtag taattctaga gctaatacgt gcaacaaacc ccgacttctg gaagggacgc"
[28] "      181 atttattaga taaaaggtcg acgcgggctt tgcccgatgc tctgatgatt catgataact"
[29] "      241 tgacggatcg cacggccatc gtgccggcga cgcatcattc aaatttctgc cctatcaact"
[30] "      301 ttcgatggta ggatagtggc ctaccatggt gtttacgggt aacggagaat tagggttcga"
[31] "      361 ttccggagag ggagcctgag aaacggctac cacatccaag gaaggcagca ggcgcgcaaa"
[32] "      421 ttacccaatc ctgacacggg gaggtagtga caataaataa caatactggg ctctttgagt"
```

```
[33] "     481 ctggtaattg gaatgagtac aatctaaatc ccttaacgag gatccattgg agggcaagtc"
[34] "     541 tggtgccagc agccgcggta attccagctc caatagcgta tatttaagtt gttgcagtta"
[35] "     601 aaaagctcgt agttggacct tggggttgggt cgaccggtcc gcctttggtg tgcaccgatc"
[36] "     661 ggctcgtccc ttctaccggc gatgcgcgcc tggccttaat tggccgggtc gttcctccgg"
[37] "     721 tgctgttact ttgaagaaat tagagtgctc aaagcaagca ttagctatga atacattagc"
[38] "     781 atgggataac attataggat tccgatccta ttatgttggc cttcgggatc ggagtaatga"
[39] "     841 ttaacaggga cagtcggggg cattcgtatt tcatagtcag aggtgaaatt cttggattta"
[40] "     901 tgaaagacga acaactgcga aagcatttgc caaggatgtt ttcattaatc aagaacgaaa"
[41] "     961 gttgggggct cgaagacgat cagataccgt cctagtctca accataaacg atgccgacca"
[42] "    1021 gggatcagcg aatgttactt ttaggacttc gctggcacct tatgagaaat caaagttttt"
[43] "    1081 gggttccggg gggagtatgg tcgcaaggct gaaacttaaa ggaattgacg gaagggcacc"
[44] "    1141 accaggagtg gagcctgcgg cttaatttga ctcaacacgg ggaaacttac caggtccaga"
[45] "    1201 catagtaagg attgacagac tgagagctct ttcttgattc tatgggtggt ggtgcatggc"
[46] "    1261 cgttcttagt tggtggagcg atttgtctgg ttaattccgt taacgaacga gacctcagcc"
[47] "    1321 tgctaactag ctatgtggag gtacacctcc acggccagct tcttagaggg actatggccg"
[48] "    1381 cttaggccaa ggaagtttga ggcaataaca ggtctgtgat gcccttagat gttctgggcc"
[49] "    1441 gcacgcgcgc tacactgatg tattcaacga gtctatagcc ttggccgaca ggcccgggta"
[50] "    1501 atctttgaaa tttcatcgtg atggggatag atcattgcaa ttgttggtct tcaacgagga"
[51] "    1561 attcctagta agcgcgagtc atcagctcgc gttgactacg tccctgccct ttgtacacac"
[52] "    1621 cgcccgtcgc tcctaccgat tgaatggtcc ggtgaaatgt tcggatcgcg gcgacgtggg"
[53] "    1681 cgcttcgtcg ccgacgacgt cgcgagaagt ccattgaacc ttatcattta gaggaaggag"
[54] "//"
```

Notice that each line is read in as a separate cell in a vector, with sequences beginning with a number ending with 1. We can take advantage of this to extract just the sequence data

7.12 Challenge

Before we move on, try to do the following:

1. Isolate only the rows containing DNA sequences. This should include

   a. Removing all of the characters that are not a, t, g, or c.
   b. Combining separate cells/lines into a single string. You can do this with using the `paste()` function with the `collapse=""` parameter

7.13. SOLUTIONS

2. Convert lower-case to upper-case. To do this, you can use:

```
gsub("([actg])","\\U\\1",Seq,perl=T)
```

The \\U\\\1 means "paste brackets as upper-case", and is only available as a **Perl** command, which is accessible in gsub() with the perl=T parameter.

3. Replace start codons (ATG) with -->START-->ATG

4. Insert >--STOP--| after any stop codons (TAA or TAG or TGA).

Take the time to struggle with this and try different combinations until you find a way through. The more you struggle, the faster you will learn.

A cool thing about regular expressions is that there is rarely a single right answer, especially for complicated problems. When you are ready, Continue on to see one possible solution.

7.13 Solutions

7.13 Transects

Subset only transect A for the first 3 species:

```
Transect[1:3,grep("A",names(Transect))]
```

```
          A1          A2         A3
1 -0.4198256 -0.13312498 -1.410915
2  0.9787430  1.19501106 -1.283054
3 -1.3156918  0.09378044  0.733655
```

Subset the data for the species "d".

```
Transect[grep("d",Transect$Species),]
```

```
  Species        A1         A2       A3        B1        B2
4       d 0.4451767 -0.3385702 1.233535 0.7391345 0.4087561
         B3         C1        C2        C3
4 0.7041051 -0.2428876 0.8681144 -1.786355
```

7.13 Genbank

First, isolate the DNA sequence:

1. Use .* with () to delete everything before the DNA sequence

```
Seq<-gsub(".*(1 [gatc])","",Lythrum_18S)
```

2. Use the .* and space with + to eliminate all text before the sequence

```
Seq<-gsub(".*ORIGIN +","",paste(Seq,collapse=""))
```

3. Eliminate spaces and the two // at the end

```
Seq<-gsub(" |//","",Seq)
```

4. Capital letters look nicer, but requires a PERL qualifier \\U in the replacement string. This is not standard in R regular expressions, but can be accessed with perl=T.

```
Seq<-gsub("([actg])","\\U\\1",Seq,perl=T)
```

Now that we have our formatted DNA sequence string, we can highlight start codons.

```
ORFs<-gsub("ATG","\n <ATG>-->START-->",Seq)
```

And stop codons too.

```
ORFs<-gsub("(TAA|TAG|TGA)","<\\1>--STOP--| \n",Seq)
```

Note the addition of the newline character (\n) to make the output more readable. To see the final modified string, use the `cat()` function to print the string directly, rather than creating an object containing the string:

```
cat(ORFs)
```

(*Output not shown*)

7.14 More Exercises

Here are some more exercises to practice your skills. No solutions are given, but you will know if you are correct if you get the desired output.

1. **Email Spammer**: Consider a vector of email addresses scraped from the internet.

   - robert 'dot' colautti 'at' queensu 'dot' ca
   - chris.eckert[at]queensu.ca

- lonnie.aarssen at queensu.ca

Use regular expressions to convert all email addresses to the standard format: name@queensu.ca

2. **Genetic Simulation**: Start by creating a random sequence of DNA. Think way back to the *R Fundamentals* Chapter to find a function that can randomly sample from 4 base pairs to create vector of 1,000 bases.

Once you have your vector, try collapsing it into a single element (i.e., a string with 1,000 characters).

Hint: You can do this with the `paste()` command. You just need one special parameter.

Now, try each of the following challenges:

- Replace T with U.
- Find all start codons (AUG) and stop codons (UAA, UAG, UGA).
- Find all open reading frames (hint: consider each sequence beginning with AUG and ending with a stop codon; how do you know if both sequences are in the same reading frame?).
- Count the length (number of bases) for all open reading frames.

3 Regex Golf

Have fun! *https://alf.nu/RegexGolf*

4. **More online examples**

http://regex.sketchengine.co.uk/extra_regexps.html

Chapter 8

Data Science

8.1 Overview

Data Science is a relatively new field of study that merges **computer science** and **statistics** to answer questions in other domains (e.g. business, medicine, biology, psychology). Data Science as a discipline has grown in popularity in response to the rapid rate of increase in data collection and publication.

Data Science often involves 'Big Data', which doesn't have a strict quantitative definition but will usually have one or more of the following characteristics:

1. High **Volume** – large file sizes with may observations.
2. Wide **Variety** – many different types of data.
3. High **Velocity** – data accumulates at a high rate.
4. Compromised **Veracity** – data quality issues must be addressed otherwise downstream analyses will be compromised.

Question: What are some examples of 'big data' in Biology?

Answer: There are many types of biological data that could be listed, but medical records, remote sensing data, and 'omics data are common examples of 'big data' in biology.

In biology, it can be helpful to think of Data Science as a continuous life-cycle with multiple stages:

8.1 Data Science Life-Cycle

1. **Hypothesize** – Make initial observations about the natural world, or insights from other data, that lead to testable hypotheses. Your core biology training is crucial here.

2. **Collect** – This may involve taking measurements yourself, manually entering data that is not yet in electronic format, requesting data from authors of published studies, or importing data from online sources. Collecting data is a crucial step that is often done poorly.

3. **Correct** – Investigate the data for quality assurance, to identify and fix potential errors. Start to visualize the data to look for outliers, odd frequency distributions, or nonsensical relationships among variables.

4. **Explore** – Try to understand the data, where they come from, and potential limitations on their use. Continue visualizing data; this may cause you to modify your hypotheses slightly.

5. **Model** – Now that hypotheses are clearly defined, apply statistical tests of their validity.

6. **Report** – Use visualizations along with the results of your statistical tests to summarise your findings.

7. **Repeat** – Return to step 1.

In this chapter, we focus mainly on coding in R for steps 2, 3, and 6. Step 5 requires a firm understanding of statistics, which is the focus of the book *R STATS Crash Course for Biologists*. Visualizations in Steps 3 & 4 were covered in earlier chapters.. Step 1 requires a good understanding of the study system, as covered in a typical university degree in the biological sciences.

Data collection and management are crucial steps in the Data Science Life-Cycle. Read the baRcodeR paper by Wu et al (2022) (*https://doi.org/10.1111/2041-210X.13405*). called *baRcodeR with PyTrackDat: Open-source labelling and tracking of biological samples for repeatable science*. Pay particular attention to the '*Data Standards*' section. The *baRcodeR* and *PyTrackDat* programs and their application to current projects may also be of interest.

8.2 Setup

The **tidyverse** library is a set of packages created by the same developers responsible for R Studio. There are a number of packages and functions that improve on base R. The name is based on the idea of living in a data universe that is neat and *tidy*.

The book *R for Data Science* (*http://shop.oreilly.com/product/0636920034407.do*) by Hadley Wickham & Garrett Grolemund is an excellent resource for learning about the tidyverse. In this chapter we'll touch on a few of the main packages and functions.

> **Protip** In general, any book by Hadley Wickham that you come across is worth reading if you want to be proficient in R.

While we wait for the `tidyverse` to install, let's consider a few general principles of data science.

8.3 2D Data Wrangling

The `dplyr` library in R has many useful features for importing and re-organizing your data for steps 2, 3 and 4 in the Data Science Life-Cycle.

`library(dplyr)`

> **Note**: This error message informs us that the `dplyr` package uses function or parameter names that are the same as other base or stats packages in R. These base/stats functions are 'masked' meaning that when you run one (e.g. `filter`) then R will run the `dplyr` version rather than the stats version.

`library(tidyr)`

We'll work with our `FallopiaData.csv` data set, and remind ourselves of the structure of the data

8.3 `tibbles` and `readr()`

We looked at `data.frame` objects in the first chapter as an expansion of *matrices* with a few additional features like column and row names. A `tibble` is the `tidyverse` version of the `data.frame` object and includes a few more useful features. To import a dataset to a `tibble` instead of a `data.frame` object, we use `read_csv` instead of `read.csv`.

8.3. 2D DATA WRANGLING

```
library(tidyverse)
```

```
-- Attaching packages ----------------- tidyverse 1.3.2 --
v tibble   3.1.8     v stringr  1.4.1
v readr    2.1.3     v forcats  0.5.2
v purrr    0.3.5
-- Conflicts -------------------- tidyverse_conflicts() --
x gridExtra::combine() masks dplyr::combine()
x dplyr::filter()      masks stats::filter()
x dplyr::lag()         masks stats::lag()
x readr::parse_date()  masks curl::parse_date()
```

```
Fallo<-read_csv(
   "https://colauttilab.github.io/RCrashCourse/FallopiaData.csv")
str(Fallo)
```

```
Rows: 123 Columns: 13
-- Column specification ----------------------------------
Delimiter: ","
chr  (3): Scenario, Nutrients, Taxon
dbl (10): PotNum, Symphytum, Silene, Urtica, Geranium, G...

i Use `spec()` to retrieve the full column specification for this data.
i Specify the column types or set `show_col_types = FALSE` to quiet this message.

spc_tbl_ [123 x 13] (S3: spec_tbl_df/tbl_df/tbl/data.frame)
 $ PotNum      : num [1:123] 1 2 3 5 6 7 8 9 10 11 ...
 $ Scenario    : chr [1:123] "low" "low" "low" "low" ...
 $ Nutrients   : chr [1:123] "low" "low" "low" "low" ...
 $ Taxon       : chr [1:123] "japon" "japon" "japon" "japon" ...
 $ Symphytum   : num [1:123] 9.81 8.64 2.65 1.44 9.15 ...
 $ Silene      : num [1:123] 36.4 29.6 36 21.4 23.9 ...
 $ Urtica      : num [1:123] 16.08 5.59 17.09 12.39 5.19 ...
 $ Geranium    : num [1:123] 4.68 5.75 5.13 5.37 0 9.05 3.51 9.64 7.3 6.36 ...
 $ Geum        : num [1:123] 0.12 0.55 0.09 0.31 0.17 0.97 0.4 0.01 0.47 0.33 ...
 $ All_Natives : num [1:123] 67 50.2 61 40.9 38.4 ...
 $ Fallopia    : num [1:123] 0.01 0.04 0.09 0.77 3.4 0.54 2.05 0.26 0 0 ...
 $ Total       : num [1:123] 67.1 50.2 61.1 41.7 41.8 ...
 $ Pct_Fallopia: num [1:123] 0.01 0.08 0.15 1.85 8.13 1.12 3.7 0.61 0 0 ...
```

```
- attr(*, "spec")=
  .. cols(
  ..   PotNum = col_double(),
  ..   Scenario = col_character(),
  ..   Nutrients = col_character(),
  ..   Taxon = col_character(),
  ..   Symphytum = col_double(),
  ..   Silene = col_double(),
  ..   Urtica = col_double(),
  ..   Geranium = col_double(),
  ..   Geum = col_double(),
  ..   All_Natives = col_double(),
  ..   Fallopia = col_double(),
  ..   Total = col_double(),
  ..   Pct_Fallopia = col_double()
  .. )
- attr(*, "problems")=<externalptr>
```

This file is an example of a 2-dimensional data set, which is common in biology. 2D datasets have the familiar row x column layout used by spreadsheet programs like Microsoft Excel or Google Sheets. There are some exceptions, but data in this format should typically follows 3 rules:

1. Each cell contains a single value
2. Each variable must have its own column
3. Each observation must have its own row

Making sure your data are arranged this way will usually make it much easier to work with.

8.3 `filter()` Rows

The `filter()` function will subset observations based on values of interest within particular columns of our dataset. For example, we may want to *filter* the rows (i.e. pots) that had at least 70 g of total biomass.

8.3. 2D DATA WRANGLING

```
Pot1<-filter(Fallo,Total >= 70)
head(Pot1)
```

```
# A tibble: 6 x 13
  PotNum Scena~1 Nutri~2 Taxon Symph~3 Silene Urtica Geran~4
   <dbl> <chr>   <chr>   <chr>   <dbl>  <dbl>  <dbl>   <dbl>
1     60 high    high    bohem    7.77   51.4   5.13    10.1
2     67 gradual high    japon    2.92   25.2  19.1     23.1
3     70 gradual high    japon   10.1    47.0  18.6      0.64
4     86 gradual high    bohem    2.93   60.9   4.11     6.67
5     95 extreme high    japon    4.92   25.9  40.3      4.92
6    103 extreme high    japon    6.92   49.4   0       10.3
# ... with 5 more variables: Geum <dbl>, All_Natives <dbl>,
#   Fallopia <dbl>, Total <dbl>, Pct_Fallopia <dbl>, and
#   abbreviated variable names 1: Scenario, 2: Nutrients,
#   3: Symphytum, 4: Geranium
```

8.3 rename() Columns

There are different options to change the names of columns in your data. In base R you can use the names() function with the square bracket index [ ]:

```
X<-Fallo
names(X)
```

```
 [1] "PotNum"       "Scenario"     "Nutrients"
 [4] "Taxon"        "Symphytum"    "Silene"
 [7] "Urtica"       "Geranium"     "Geum"
[10] "All_Natives"  "Fallopia"     "Total"
[13] "Pct_Fallopia"
```

```
names(X)[12]<-"Total_Biomass"
names(X)
```

```
 [1] "PotNum"       "Scenario"     "Nutrients"
 [4] "Taxon"        "Symphytum"    "Silene"
 [7] "Urtica"       "Geranium"     "Geum"
[10] "All_Natives"  "Fallopia"     "Total_Biomass"
[13] "Pct_Fallopia"
```

There is also a simple dplyr function to do this:

```
X<-rename(Fallo, Total_Biomass = Total)
names(X)
```

```
 [1] "PotNum"       "Scenario"     "Nutrients"
 [4] "Taxon"        "Symphytum"    "Silene"
 [7] "Urtica"       "Geranium"     "Geum"
[10] "All_Natives"  "Fallopia"     "Total_Biomass"
[13] "Pct_Fallopia"
```

8.3 arrange() Rows

Use the arrange() function to sort the *rows* of your data based on the values in one or more *columns*. For example, let's re-arrange our FallopiaData.csv dataset based on Taxon (a string denoting the species of Fallopia used) and Total (a float denoting the total biomass in each pot).

```
X<-arrange(Fallo, Taxon, Total)
head(X)
```

```
# A tibble: 6 x 13
  PotNum Scena~1 Nutri~2 Taxon Symph~3 Silene Urtica Geran~4
```

8.3. 2D DATA WRANGLING

```
     <dbl> <chr>   <chr> <chr>  <dbl> <dbl> <dbl> <dbl>
1       26 low     low   bohem  13.2   18.1  0     0
2       17 low     low   bohem   4.9   29.5  1.36  0
3       80 gradual high  bohem  11.9   17.2  8.92  0.94
4       18 low     low   bohem   3.51  27.6  8.14  3.81
5       28 low     low   bohem  10.6   18.8  7.19  6.73
6       22 low     low   bohem   0.76  22.7  9.85 10.6
# ... with 5 more variables: Geum <dbl>, All_Natives <dbl>,
#   Fallopia <dbl>, Total <dbl>, Pct_Fallopia <dbl>, and
#   abbreviated variable names 1: Scenario, 2: Nutrients,
#   3: Symphytum, 4: Geranium
```

use the desc() function with arrange() to reverse and sort in *descending* order. We can sort by multiple columns in the by_group= parameter.

```
X<-arrange(Fallo, by_group=Taxon, desc(Silene))
head(X)
```

```
# A tibble: 6 x 13
  PotNum Scena~1 Nutri~2 Taxon Symph~3 Silene Urtica Geran~4
   <dbl> <chr>   <chr>   <chr>   <dbl>  <dbl>  <dbl>   <dbl>
1     86 gradual high    bohem    2.93   60.9   4.11    6.67
2     53 high    high    bohem    7.05   56.3   1.14    4.07
3     60 high    high    bohem    7.77   51.4   5.13   10.1
4     50 high    high    bohem    8.52   44.6   1.27    9.45
5     79 gradual high    bohem   11.0    44.6   1.56    0.03
6    107 extreme high    bohem    0      43.8   8.1     8.18
# ... with 5 more variables: Geum <dbl>, All_Natives <dbl>,
#   Fallopia <dbl>, Total <dbl>, Pct_Fallopia <dbl>, and
#   abbreviated variable names 1: Scenario, 2: Nutrients,
#   3: Symphytum, 4: Geranium
```

8.3 `select()` Columns

The `select()` function can be used to select a subset of columns (i.e. variables) from your data.

Suppose we only want to look at total biomass, but keep all the treatment columns:

```
X<-select(Fallo, PotNum, Scenario, Nutrients, Taxon, Total)
head(X)
```

```
# A tibble: 6 x 5
  PotNum Scenario Nutrients Taxon Total
   <dbl> <chr>    <chr>     <chr> <dbl>
1      1 low      low       japon  67.1
2      2 low      low       japon  50.2
3      3 low      low       japon  61.1
4      5 low      low       japon  41.7
5      6 low      low       japon  41.8
6      7 low      low       japon  48.3
```

You can also use the colon `:` to select a range of columns:

```
X<-select(Fallo, PotNum:Taxon, Total)
head(X)
```

```
# A tibble: 6 x 5
  PotNum Scenario Nutrients Taxon Total
   <dbl> <chr>    <chr>     <chr> <dbl>
1      1 low      low       japon  67.1
2      2 low      low       japon  50.2
3      3 low      low       japon  61.1
4      5 low      low       japon  41.7
5      6 low      low       japon  41.8
6      7 low      low       japon  48.3
```

8.3. 2D DATA WRANGLING

Exclude columns with -

```
X<-select(Fallo, -PotNum:Taxon, -Total)
```

```
Warning in x:y: numerical expression has 12 elements: only
the first used
```

> Oops, what generated that warning? Take a careful look at the error message and see if you can figure it out.

The problem is we are using the range of columns between PotNum and Taxon, but in one case we are excluding and the other we are including. We need to be consistent:

```
X<-select(Fallo, -PotNum:-Taxon, Total)
head(X)
```

```
# A tibble: 6 x 9
  Symphy~1 Silene Urtica Geran~2  Geum All_N~3 Fallo~4 Total
     <dbl>  <dbl>  <dbl>   <dbl> <dbl>   <dbl>   <dbl> <dbl>
1     9.81   36.4   16.1    4.68  0.12    67.0    0.01  67.1
2     8.64   29.6   5.59    5.75  0.55    50.2    0.04  50.2
3     2.65   36.0   17.1    5.13  0.09    61.0    0.09  61.1
4     1.44   21.4   12.4    5.37  0.31    40.9    0.77  41.7
5     9.15   23.9   5.19       0  0.17    38.4     3.4  41.8
6     6.31   24.4      7    9.05  0.97    47.7    0.54  48.3
# ... with 1 more variable: Pct_Fallopia <dbl>, and
#   abbreviated variable names 1: Symphytum, 2: Geranium,
#   3: All_Natives, 4: Fallopia
```

Or a bit more clear:

```
X<-select(Fallo, -(PotNum:Taxon), Scenario)
head(X)
```

```
# A tibble: 6 x 10
  Symphy~1 Silene Urtica Geran~2  Geum All_N~3 Fallo~4 Total
     <dbl>  <dbl>  <dbl>   <dbl> <dbl>   <dbl>   <dbl> <dbl>
1     9.81   36.4   16.1    4.68  0.12    67.0    0.01  67.1
2     8.64   29.6   5.59    5.75  0.55    50.2    0.04  50.2
3     2.65   36.0   17.1    5.13  0.09    61.0    0.09  61.1
4     1.44   21.4   12.4    5.37  0.31    40.9    0.77  41.7
5     9.15   23.9   5.19    0     0.17    38.4    3.4   41.8
6     6.31   24.4   7       9.05  0.97    47.7    0.54  48.3
# ... with 2 more variables: Pct_Fallopia <dbl>,
#   Scenario <chr>, and abbreviated variable names
#   1: Symphytum, 2: Geranium, 3: All_Natives, 4: Fallopia
```

8.3 everything()

Use the everything() function with select() to rearrange your columns without losing any:

```
X<-select(Fallo, Taxon, Scenario, Nutrients, PotNum,
        Pct_Fallopia, everything())
head(X)
```

```
# A tibble: 6 x 13
  Taxon Scena~1 Nutri~2 PotNum Pct_F~3 Symph~4 Silene Urtica
  <chr> <chr>   <chr>    <dbl>   <dbl>   <dbl>  <dbl>  <dbl>
1 japon low     low          1    0.01    9.81   36.4   16.1
2 japon low     low          2    0.08    8.64   29.6   5.59
3 japon low     low          3    0.15    2.65   36.0   17.1
4 japon low     low          5    1.85    1.44   21.4   12.4
5 japon low     low          6    8.13    9.15   23.9   5.19
```

8.3. 2D DATA WRANGLING

```
6 japon low      low              7    1.12    6.31   24.4      7
# ... with 5 more variables: Geranium <dbl>, Geum <dbl>,
#   All_Natives <dbl>, Fallopia <dbl>, Total <dbl>, and
#   abbreviated variable names 1: Scenario, 2: Nutrients,
#   3: Pct_Fallopia, 4: Symphytum
```

8.3 mutate() Columns

Suppose we want to make a new column to our data frame or tibble. For example, we calculate the sum of biomass of Urtica and Geranium only. In base R, we could use $ to select the column from the data frame.

```
X<-Fallo
X$UrtSil<-X$Urtica+X$Silene
```

In the dplyr package we can use the mutate() function.

```
X<-mutate(Fallo, UrtSil = Urtica + Silene)
head(X)
```

```
# A tibble: 6 x 14
  PotNum Scena~1 Nutri~2 Taxon Symph~3 Silene Urtica Geran~4
   <dbl> <chr>   <chr>   <chr>   <dbl>  <dbl>  <dbl>   <dbl>
1      1 low     low     japon    9.81   36.4   16.1    4.68
2      2 low     low     japon    8.64   29.6    5.59   5.75
3      3 low     low     japon    2.65   36.0   17.1    5.13
4      5 low     low     japon    1.44   21.4   12.4    5.37
5      6 low     low     japon    9.15   23.9    5.19   0
6      7 low     low     japon    6.31   24.4    7      9.05
# ... with 6 more variables: Geum <dbl>, All_Natives <dbl>,
#   Fallopia <dbl>, Total <dbl>, Pct_Fallopia <dbl>,
#   UrtSil <dbl>, and abbreviated variable names
#   1: Scenario, 2: Nutrients, 3: Symphytum, 4: Geranium
```

This is a lot more readable, especially when you have complicated equations or you want to add many of new columns.

> **Question**: What if you only wanted to retain the new columns and delete everything else? Try it.

Which functions did you use?

8.3 `transmute()` Columns

The `transmute()` functions acts as a combination of `mutate()` + `select()`

```
X<-transmute(Fallo, UrtSil = Urtica + Silene)
head(X)
```

```
# A tibble: 6 x 1
   UrtSil
    <dbl>
1   52.4
2   35.2
3   53.1
4   33.8
5   29.1
6   31.4
```

8.3 `summarise()` + `group_by()`

This can be useful for quickly summarizing your data, for example to find the mean or standard deviation based on a particular treatment or group.

```
TrtGrp<-group_by(Fallo,Taxon,Scenario,Nutrients)
summarise(TrtGrp, Mean=mean(Total), SD=sd(Total))
```

`summarise()` has grouped output by 'Taxon', 'Scenario'.
You can override using the `.groups` argument.

```
# A tibble: 10 x 5
# Groups:   Taxon, Scenario [10]
   Taxon Scenario     Nutrients  Mean    SD
   <chr> <chr>        <chr>      <dbl> <dbl>
 1 bohem extreme      high        58.3  7.34
 2 bohem fluctuations high        58.4  9.20
 3 bohem gradual      high        57.5  9.34
 4 bohem high         high        60.3  8.68
 5 bohem low          low         48.0  8.86
 6 japon extreme      high        57.2 10.9
 7 japon fluctuations high        56.4 13.7
 8 japon gradual      high        59.7  9.57
 9 japon high         high        56.4  8.20
10 japon low          low         52.0  8.29
```

8.3 Weighted Mean

In our dataset, the **Taxon** column shows which of two species of *Fallopia* were used in the competition experiments. We might want to take the mean total biomass for each of the two *Fallopia* species.

However, there are other factors in our experiment that may affect biomass. The *Nutrients* column tells us whether pots received high or low nutrients, and this also affects biomass:

```
X<-group_by(Fallo,Nutrients)
summarise(X, Mean=mean(Total), SD=sd(Total))
```

```
# A tibble: 2 x 3
  Nutrients  Mean    SD
  <chr>     <dbl> <dbl>
1 high       58.0  9.61
2 low        49.9  8.66
```

We can see that the Nutrients treatment had a large effect on the mean biomass of a pot. Now imagine if our sampling design is 'unbalanced'. For example, maybe we had some plant mortality or lost some plants to a tornado. If one of the two species in the Taxon column had more high-nutrient pots left over, then it would have a higher mean. BUT, what if the nutrients promoted growth? We would expect to see a difference in the mean of each Taxon group, this expected difference is just an artifact of unbalanced replication of the high nutrient treatment. We can simulate this effect by re-shuffling the species names:

```
RFallo<-Fallo
set.seed(256)
RFallo$Taxon<-rbinom(nrow(RFallo),size=1,prob=0.7)

X<-group_by(RFallo,Taxon)
summarise(X, Mean=mean(Total))
```

```
# A tibble: 2 x 2
  Taxon  Mean
  <int> <dbl>
1     0  56.1
2     1  56.5
```

In this example, the difference is less than 1% of the mean (i.e., $0.4/56.1 = 0.0071 = 0.71$), but this could be due to the imbalance offsetting biological differences between the taxa. In other scenarios it could be much larger. To fix this problem, we may want to take a

8.3. 2D DATA WRANGLING

weighted mean. In this case, we want to *weight* the mean of each taxon by the unbalanced Scenario and Treatment categories.

Step 1: Group by all three columns and calculate group means

```
X1<-group_by(RFallo,Taxon,Scenario,Nutrients)
(Y1<-summarise(X1,Mean=mean(Total)))
```

```
# A tibble: 10 x 4
# Groups:   Taxon, Scenario [10]
   Taxon Scenario     Nutrients  Mean
   <int> <chr>        <chr>      <dbl>
 1     0 extreme      high       54.7
 2     0 fluctuations high       58.4
 3     0 gradual      high       56.2
 4     0 high         high       59.8
 5     0 low          low        50.2
 6     1 extreme      high       59.1
 7     1 fluctuations high       56.8
 8     1 gradual      high       59.9
 9     1 high         high       57.6
10     1 low          low        49.8
```

Now we have a single value for each of the subgroups. Instead of over-representing pots from Taxon=1, we have equal weightings of all three groups.

Step 2: Take the output from Step 1 and calculate the *mean of means* for the column of interest.

```
X2<-group_by(Y1,Taxon)
Y2<-summarise(X2, Mean=mean(Mean))
arrange(Y2,desc(Mean))
```

```
# A tibble: 2 x 2
```

```
  Taxon  Mean
  <int>  <dbl>
1   1    56.6
2   0    55.8
```

Now there is more than a 2.1 difference (i.e., $1.2/55.8 = 0.0215$) between the taxa. This is triple the effect that we calculated with unbalanced samples!

8.3 %>% (pipe)

The dplyr package includes a special command (%>%) called a *pipe*. The *pipe* is a very handy tool for complex data wrangling. It allows us to string together multiple functions and then *pipe* our data from one function to the next. In principle, this is similar to the multi-function plotting commands we made with ggplot() in earlier chapters.

The pipe is useful to combine operations without creating a whole bunch of new objects. This can save on memory use and run speed because you are not making new objects for every single function.

Pro-tip: Use Ctl + Shift + m to add a pipe quickly

For example, we can re-write the weighted mean example using pipes.

```
RFallo %>%
  group_by(Taxon,Scenario,Nutrients) %>%
  summarise(Mean=mean(Total)) %>%
  group_by(Taxon,Scenario) %>%
  summarise(Mean=mean(Mean)) %>%
  group_by(Taxon) %>%
  summarise(Mean=mean(Mean)) %>%
  arrange(desc(Mean))
```

8.3. 2D DATA WRANGLING

```
# A tibble: 2 x 2
  Taxon  Mean
  <int>  <dbl>
1   1    56.6
2   0    55.8
```

We declare the input data set in the first line, and then pipe to group_by() in line 2. The output of group_by() is then *piped to summarise(), and so on. There are two important things to note here:

1. We do not declare the data object inside of each function. The pipe command does this for us.
2. We still have to assign the output to an object if we want to retain it. For example, we might change line 1 to wtMean<-RFallo %>%. We didn't redirect the command to an object, so it simply output to your R Console.

Compared to the earlier example, it is clear that the pipe can make for cleaner, more concise code with fewer objects added to the environment. also avoids potential for bugs in our program. Imagine if we mis-spelled 'Taxon' in our second line by accidentally pressing 's' along with 'x'. Compare the output:

```
X<-group_by(Fallo,Taxon,Scenario,Nutrients)
X<-group_by(X,Tasxon,Scenario)


Error in `group_by()`:
! Must group by variables found in `.data`.
x Column `Tasxon` is not found.
```

```
X<-group_by(X,Taxon)
X<-summarise(X, Mean=mean(Total), SD=sd(Total))
arrange(X,desc(Mean))
```

```
# A tibble: 2 x 3
  Taxon  Mean    SD
  <chr>  <dbl> <dbl>
1 japon   56.4  10.4
2 bohem   56.3   9.54
```

```
Fallo %>%
  group_by(Taxon,Scenario,Nutrients) %>%
  group_by(Tasxon,Scenario) %>%
  group_by(Taxon) %>%
  summarise(Mean=mean(Total), SD=sd(Total)) %>%
  arrange(desc(Mean))
```

```
Error in `group_by()`:
! Must group by variables found in `.data`.
x Column `Tasxon` is not found.
```

In both cases we get an error, but in one case we still calculate the means and sd of the two species.

> A bug that produces no output is much less dangerous than an error that gives an output. Why?

8.4 Join datasets

The dplyr package has some handy `joing_` tools to combine data sets. There are four main ways to join data sets. I find it helpful to think of Venn diagrams when I'm trying to remember what all of these functions

8.4. JOIN DATASETS

do. You can get more information on these with the help command ?join after loading the dplyr library, but here is a quick overview. For each of these, imagine a Venn diagram with two datasets: X as a circle on the left side and Y as a circle on the right side. The rows we choose to combine from the two datasets depend on one or more identifying columns that we can define (e.g. sample ID, date, time).

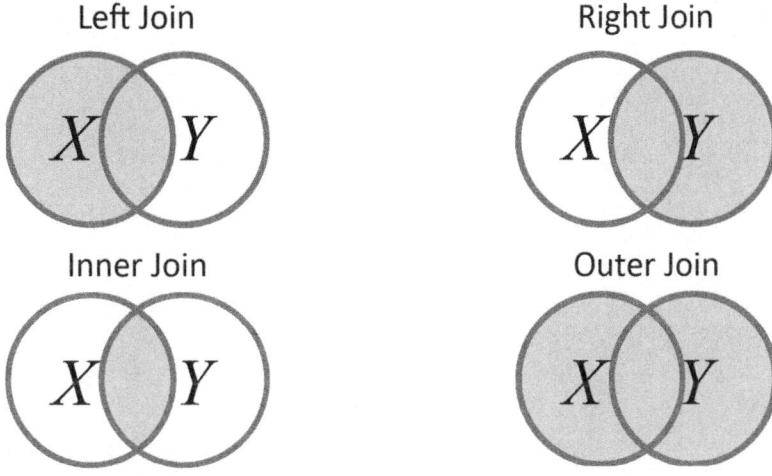

Figure 8.1: Venn diagram of datasets

1. left_join() - Keep all rows of X and add matching rows from Y. Any rows in Y that don't match X are excluded.

2. right_join() - The reverse of left_join()

3. inner_join() - Only keep rows that are common to **both** X **AND** Y. Remove any rows that don't match.

4. full_join() - Keep **any** columns that are in **either** X **OR** Y. Add NA for missing data in either of the columns for rows that don't match.

To try these out, we can create a couple of quick data sets and compare the output. For each case, note the addition of NA for missing data.

```
X<-data.frame(ID=c("A","C"),Xdat=c(1,3))
Y<-data.frame(ID=c("B","C"),Ydat=c(2,3))
X
```

```
  ID Xdat
1  A    1
2  C    3
```

```
Y
```

```
  ID Ydat
1  B    2
2  C    3
```

```
left_join(X,Y,by="ID")
```

```
  ID Xdat Ydat
1  A    1   NA
2  C    3    3
```

```
right_join(X,Y,by="ID")
```

```
  ID Xdat Ydat
1  C    3    3
2  B   NA    2
```

```
inner_join(X,Y,by="ID")
```

```
  ID Xdat Ydat
1  C    3    3
```

8.5. WIDE VS LONG DATA

```
full_join(X,Y,by="ID")
```

```
  ID Xdat Ydat
1  A    1   NA
2  C    3    3
3  B   NA    2
```

8.5 Wide vs Long data

Most of the data examples we've looked at are in the 'wide' format, where we have a single individual as a row, and multiple measurements as separate columns.

However, there are many cases where we may want to reorganize our data into the 'long' format, where each row is an individual observation. Many statistical models use this format, and it's also useful for visualizations.

Here is one very handy example that I use all the time. If we have a bunch of columns of observations and we want to generate plots quickly (e.g. frequency distributions), we can convert the data to the *long* format and then use the `facet_` functions from `ggplot2` to create separate plots for each of the original columns containing the observations.

8.5 `pivot_longer`

The `pivot_longer()` function in the `tidyr` library converts data from *wide to long*. We use the `cols=` parameter to specify the data columns, then `names_to=` specifies the column name containing the parameter, and the `cols_to=` specifies the column name of the values.

This is a bit confusing to read, but it's easier to understand if you compare the output with the `full_join` function in the previous section.

```
LongData<-full_join(X,Y,by="ID") %>%
   pivot_longer(cols=c("Xdat","Ydat"),
                names_to="Measurement",
                values_to="Value")
LongData
```

```
# A tibble: 6 x 3
   ID    Measurement Value
   <chr> <chr>       <dbl>
1  A     Xdat            1
2  A     Ydat           NA
3  C     Xdat            3
4  C     Ydat            3
5  B     Xdat           NA
6  B     Ydat            2
```

This is the *long* format. Compare to the *wide* format:

```
full_join(X,Y,by="ID")
```

```
  ID Xdat Ydat
1  A    1   NA
2  C    3    3
3  B   NA    2
```

Note how there is only one column of data values, with the Measurement column indicating which measurement the value belongs to, and the ID column is repeated for each measurement.

This is why the *long* data format is sometimes called the *repeated measures* data format.

8.5 `pivot_wider`

The `pivot_wider()` function does the reverse. This time, we specify the column that contains the values with `values_from=` and the corresponding column names with `names_from=`. This should recover the original data set:

```
WideData<-LongData %>%
  pivot_wider(values_from=Value,
              names_from=Measurement)
WideData
```

```
# A tibble: 3 x 3
  ID     Xdat  Ydat
  <chr> <dbl> <dbl>
1 A         1    NA
2 C         3     3
3 B        NA     2
```

Note the slight difference in output.

> **Question**: Can you explain why this output is different than the original?

Answer: The original was a `data.frame` object, but the output of `pivot_longer()` and `pivot_wider()`

8.6 Missing Data

So far we have worked on a pristine data set that has already been edited for errors. More often datasets will contain missing values.

8.6 NA and na.rm()

As we have already seen, the R language uses a special object NA to denote missing data.

```
Vec<-c(1,2,3,NA,5,6,7)
Vec
```

[1] 1 2 3 NA 5 6 7

When a function is run on a vector or other object containing NA, the function will often return NA or give an error message:

```
mean(Vec)
```

[1] NA

This is by design, because it is not always clear what NA means. Should these data be removed from consideration, or is it a zero to be included in calculations and statistical models? Many functions in R include an na.rm parameter that is set to FALSE by default. Setting it to true tells the function to ignore the NA (i.e., remove it from the calculation).

```
mean(Vec, na.rm=T)
```

[1] 4

8.6 NA vs 0

A common mistake students make is to put 0 for missing data. This can be a big problem when analyzing the data since the calculations are very different.

8.6. MISSING DATA

```
Vec1<-c(1,2,3,NA,5,6,7)
mean(Vec1, na.rm=T)
```

```
[1] 4
```

```
Vec2<-c(1,2,3,0,5,6,7)
mean(Vec2, na.rm=T)
```

```
[1] 3.428571
```

However, there are many cases where NA *does* represent a zero for purposes of calculation or statistical analysis. There is no simple rule to follow here. The best decision will depend on the specific details of the biological data and the assumptions of the particular calculation or statistic.

8.6 is.na()

In large datasets you might want to check for missing values. Let's simulate this in our *FallopiaData.csv* dataset.

To set up a test data frame, randomly select 10 rows and replace the value for 'Total' with NA.

> **Question**: Can you remember how to do this, from the *R Fundamentals* Chapter?

Answer: There are many ways that you could approach this. One way is to first randomly select from a vector representing the row number in our data frame. Then, replace the values of these particular observations with NA.

```
X<-sample(c(1:nrow(Fallo)),10,replace=F)
Fallo$Total[X]<-NA
Fallo$Total
```

```
  [1] 67.06 50.22 61.08 41.71 41.81 48.27 55.42    NA 53.53
 [10] 45.89 59.02 57.66 48.98 35.97 43.28 52.27    NA 44.61
 [19] 59.13 58.97 55.36 31.46 43.46 44.65 59.69 60.82 57.21
 [28] 34.09 58.57    NA 63.18 59.88 54.09 55.27 61.31 53.56
 [37] 52.66 64.71 61.06 45.34 64.20 57.50 68.55 49.55    NA
 [46] 54.06 66.60 74.82 53.71 49.75 58.45 66.06 67.01 70.41
 [55]    NA 63.43    NA 47.50 61.79 54.96 48.99 52.01    NA
 [64]    NA 42.47 46.18 62.56 54.36 69.54 75.91 56.34 64.97
 [73] 60.71 57.80 41.72 67.44 58.78 77.74 65.68 58.42 55.35
 [82] 50.28 55.04 39.56 71.07 45.23 57.20 67.70 52.46 60.86
 [91] 62.19 65.53 48.19 60.89 48.13 60.37 67.86 56.40 49.13
[100] 56.11 49.78 69.00 65.40    NA 63.08 60.93 29.54 49.12
[109] 68.73 31.90 69.88 69.48 47.88    NA 58.13 50.51 54.83
[118] 66.80 50.31 56.12 62.96 78.80 64.25
```

Use `is.na()` to check for missing values:

```
is.na(Fallo$Total)
```

```
 [1] FALSE FALSE FALSE FALSE FALSE FALSE FALSE  TRUE FALSE
[10] FALSE FALSE FALSE FALSE FALSE FALSE FALSE  TRUE FALSE
[19] FALSE FALSE FALSE FALSE FALSE FALSE FALSE FALSE FALSE
[28] FALSE FALSE  TRUE FALSE FALSE FALSE FALSE FALSE FALSE
[37] FALSE FALSE FALSE FALSE FALSE FALSE FALSE FALSE  TRUE
[46] FALSE FALSE FALSE FALSE FALSE FALSE FALSE FALSE FALSE
[55]  TRUE FALSE  TRUE FALSE FALSE FALSE FALSE FALSE  TRUE
[64]  TRUE FALSE FALSE FALSE FALSE FALSE FALSE FALSE FALSE
[73] FALSE FALSE FALSE FALSE FALSE FALSE FALSE FALSE FALSE
[82] FALSE FALSE FALSE FALSE FALSE FALSE FALSE FALSE FALSE
[91] FALSE FALSE FALSE FALSE FALSE FALSE FALSE FALSE FALSE
```

8.7. NAUGHTY DATA

```
[100] FALSE FALSE FALSE FALSE  TRUE FALSE FALSE FALSE FALSE
[109] FALSE FALSE FALSE FALSE FALSE  TRUE FALSE FALSE FALSE
[118] FALSE FALSE FALSE FALSE FALSE FALSE
```

Note that the output is a vector of True/False. Each cell corresponds to a value of 'Total' with TRUE indicating missing values. This is an example of a boolean variable, which has some handy properties in R.

First, we can use it as an index. For example, let's see which pots have missing 'Total' values:

```
Missing<-is.na(Fallo$Total)
Fallo$PotNum[Missing]
```

```
[1]   9  20  36  55  68  70  78  79 126 138
```

Another handy trick to check for missing values is to sum the vector:

```
sum(is.na(Fallo$Total))
```

```
[1] 10
```

This takes advantage of the fact that the Boolean TRUE/FALSE variable is equivalent to the binary 1/0 values. By default, R treats TRUE as 1 and FALSE as 0 if used in any mathematical operation.

8.7 Naughty Data

Naughty data contain the same information as a standard row x column (i.e. 2-dimensional) data frame but break tone or more of the following best practices.

1. Each cell contains a single value
2. Each variable must have its own column
3. Each observation must have its own row

Naughty data are very common in biology, this often occurs when either the collector(s) were not familiar with these best practices, or decisions are made to favour human readibility and usability rather than computer interpretability.

(A)

Date: July 1, 1984		
Observer: Walter Kovacs		
ID	Length	Yield
38681	*80*	DNA: 100
		RNA: ??
10034	**0.1**	DNA: 122
		RNA: none
80260	*19*	DNA: 88
		RNA: 72

NOTE: italics=cm, bold=m

Date: Oct 1, 1992		
Observer: Reggie Long		
ID	Length	Yield
10545	1m	DNA: 50
		RNA: 10
75262	88cm	DNA: 61
		RNA: 40
21221	0.9m	DNA: 44
		RNA: 36

(B)

ID	Date	Observer	Length	Len_metric	Nucleotide	Yield
38681	1984-07-01	Walter Kovacs	80	cm	DNA	100
38681	1984-07-01	Walter Kovacs	80	cm	RNA	NA
10034	1984-07-01	Walter Kovacs	0.1	m	DNA	122
10034	1984-07-01	Walter Kovacs	0.1	m	RNA	0
80260	1984-07-01	Walter Kovacs	19	cm	DNA	88
80260	1984-07-01	Walter Kovacs	19	cm	RNA	72
10545	1992-10-01	Reggie Long	1	m	DNA	50
10545	1992-10-01	Reggie Long	1	m	RNA	10
75262	1992-10-01	Reggie Long	88	cm	DNA	61
75262	1992-10-01	Reggie Long	88	cm	RNA	40
21221	1992-10-01	Reggie Long	0.9	m	DNA	44
21221	1992-10-01	Reggie Long	0.9	m	RNA	36

Figure 2. Example of (A) common errors in data management and (B) corresponding rearrangement of the same data to simplify reproducible data wrangling and analysis. Note that colours are added to show link between data in A and B and do not appear in the final text-based file (e.g. TXT, CSV, TSV).

Figure 8.2: Examples of Naughty Data from Wu *et al.* (2022)

Naughty data can be very time consuming to fix, but regular expressions can make this a bit easier, as discussed in the *Regular Expressions* Chapter.

8.8 Dates

As biologists, we often work with dates or time. We may want to analyze the date a sample was collected, or a measurement was taken. But dates are often encoded in formats that don't fit neatly into the usual data types that we've worked with so far. The lubridate package provides a convenient framework for switching between human-readable dates and mathematical relationships among them – for example, the number of days, minutes, or seconds between two time points.

```
library(lubridate)
```

```
Loading required package: timechange
```

```
Attaching package: 'lubridate'
```

```
The following objects are masked from 'package:base':

    date, intersect, setdiff, union
```

It can be convenient to automatically include the current date or time, especially when you are producing reports in R.

We can get the date as a date object in the *year-month-day* format,

```
today()
```

```
[1] "2022-12-26"
```

And we can get the date along with the time as a datetime object, in the *year-month-day hour:minute:second timezone* format.

```
now()
```

```
[1] "2022-12-26 13:58:03 EST"
```

This is an important distinction, because the `datetime` object extends on the `date` object to include hours, minutes, seconds, and time zone.

8.8 Run Time

We can use a `datetime` object to track the *run time* of our code. The **run time** is how long it takes to run a particular program. We first create an object to store the computer clock before the program runs, then we subtract that object from the computer clock after the program finishes:

```
Before<-now()
```

```
for(i in 1:1000){
  rpois(10,lambda=10)^rnorm(10)
}
```

```
now()-Before
```

```
Time difference of 0.01442814 secs
```

Note that your run time may be different, depending on the specifications of your computer and other processes that are running on it.

This technique can be useful for estimating the run time of a large program. Specifically, we run a fraction of the data and/or loop iterations and multiply to get an estimate the run time for the full data set.

Question: If you add a print() function to print out the result in each iteration of the for loop, how much does this slow down the run time?

Answer: Try it to find out!

For a small program like this, it's not a problem to ad a few fractions of a second. But this can scale up exponentially in 'big data' applications. For example, imagine you are sequencing a human genome at 100× coverage, which might typically involve one a billion sequences, each 300 bases long. You want to write a program to assemble all of the sequences into a full genome. How much time would it add to your program run time if you add a 1 millisecond step for each sequence, or each base pair in the experiment?

For each sequence, it would be $10^9 sequences \times 0.001s = 10^6 s$, or about 11.5 days! For each base pair, multiply by 300 – it would take more than 10 years!

These are problems of scale that don't matter so much when you are starting out, but developing an efficiency mindset early, it can pay off when we move on to bigger data projects.

Next, we'll explore the date and datetime objects in more detail.

8.8 date Objects

Human-readable dates come in many different forms, which we can encode as strings. Here are some examples that we might see for encoding the end-date of the most recent 5,126-year-long Mesoamerican calendar used by the ancient Maya civilization:

```
Date2<-"2012-12-21"
Date3<-"21.12.2012"
Date4<-"Dec 21, 2012"
Date5<-"21 December, 2012"
```

The `lubridate` package has a number of related functions that correspond to the order of entry – d for day, m for month, and y for year:

ymd(Date2)

```
[1] "2012-12-21"
```

dmy(Date3)

```
[1] "2012-12-21"
```

mdy(Date4)

```
[1] "2012-12-21"
```

dmy(Date5)

```
[1] "2012-12-21"
```

Notice the flexibility here! Some dates have dashes, dots, spaces, and commas! Yet, all are easily converted to a common object type. On the surface, these objects look like simple strings, but compare the structure of the date object with its original input string:

str(Date2)

```
 chr "2012-12-21"
```

8.8. DATES

```
str(ymd(Date2))
```

```
 Date[1:1], format: "2012-12-21"
```

Notice how the first is a simple `chr` character object, whereas the second is a `Date` object. The date object can be treated as a numeric variable, that outputs as a readable date. For example, what if we want to know what the date 90 days before or after?

```
c(ymd(Date2)-90,ymd(Date2)+90)
```

```
[1] "2012-09-22" "2013-03-21"
```

When using `date` objects, R will even account for the different number of days in each month, and adjust for leap years!

As with any function in R, we can apply `ymd()` to a column in a data frame or any other vector of strings:

```
DateVec<-c("2012-12-1","2012-12-20",
         "2012-12-23", "2012-11-05")
ymd(DateVec)
```

```
[1] "2012-12-01" "2012-12-20" "2012-12-23" "2012-11-05"
```

The elements must have the have the same order as the function, but surprisingly they don't have to have the same format:

```
MixVec<-c("2012-12-11","12-20-2012",
         "2012, December 21", "2013, Nov-11")
ymd(MixVec)
```

```
Warning: 1 failed to parse.

[1] "2012-12-11" NA              "2012-12-21" "2013-11-11"
```

Note the warning and the resulting output. The first, third, and fourth elements are converted, even though they have different formats. The second element is replaced with NA because the order is not year, month, day, as required by ymd().

Because these objects are numeric, we can also use them for plotting:

```
ggplot() +
  geom_histogram(aes(ymd(DateVec)),
                 binwidth=1)
```

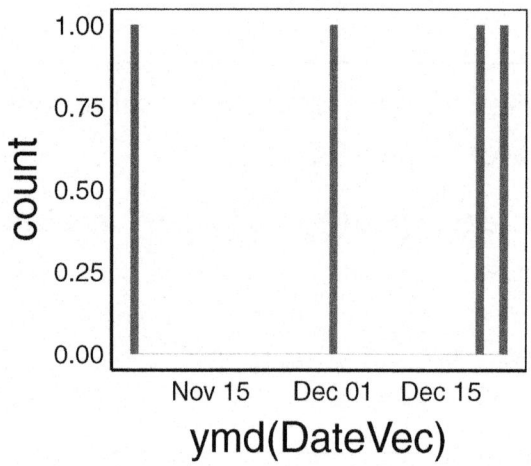

Question: What do you notice about the x-axis? How would it differ if we used strings instead?

Answer: Try to run the above with DateVec as a string instead of date.

You should get an error, because you can't bin values of a character variable to generate a frequency histogram. This shows us that R is able to treat the date object as a continuous variable, spacing the bins based on the number of days separating them.

8.8 `datetime`

The `datetime` object adds a time element to the date. Just as there are different functions to translate different date formats, there are different `datetime` functions. Each `datetime` function starts with one of the three `date` functions we used earlier, but then adds time elements after an underscore _ to define nine different functions. Here are a few examples:

```
mdy_h("Dec 21, 2012 -- 10am")
```

```
[1] "2012-12-21 10:00:00 UTC"
```

```
ymd_hm("2012-12-21, 08:30")
```

```
[1] "2012-12-21 08:30:00 UTC"
```

```
dmy_hms("21 December, 2012; 11:59:59")
```

```
[1] "2012-12-21 11:59:59 UTC"
```

8.8 Extracting Components

We can extract components from `date` and `datetime` objects. For example, we can extract only the year:

```
year(Date2)
```

```
[1] 2012
```

or the month:

```
month(Date2)
```

```
[1] 12
```

We have several options for days.

First, we can get the day of the year, also known as the Julian day:

```
yday(Date2)
```

```
[1] 356
```

Or the day of the month

```
mday(Date2)
```

```
[1] 21
```

or the day of the week (rather than month) using wday()

```
wday(Date2)
```

```
[1] 6
```

We can use the `label=T` parameter rather than number, to get the name of the specific month or day, rather than the number:

```
month(Date2, label=T)
```

```
[1] Dec
12 Levels: Jan < Feb < Mar < Apr < May < Jun < ... < Dec
```

```
wday(Date2, label=T)
```

```
[1] Fri
Levels: Sun < Mon < Tue < Wed < Thu < Fri < Sat
```

8.8 Categorical dates

We've seen above how dates are more like `numeric` objects rather than `strings`, but we can also treat dates as categorical data.

One example that is becoming more and more common in biology is the analysis of data from data loggers, which automatically save observations over time. Think of climate stations that measure temperature and precipitation as a common example. Another example might be location information of a study organism using image analysis or PIT tags (i.e. Passive Integrated Transponders).

In many cases, the timescale of collection is much shorter than what we need for our analysis. We end up with too much data! Luckily, `dplyr` with `lubridate` offer a great combination for summarizing these data.

8.8.5.1 `lubridate` + `dplyr`

Here's an example that takes advantage that takes advantage of our ability to treat components of `datetime` objects as categorical variables as well as continuous variables.

Imagine we have observations taken every minute, and we just want to calculate the average for each hour . To see how to do this, we will generate a toy data set in R using the tibble function, and then assigning column names: `DayTime` for the day-time object and `Obs` for the observation (i.e. measurement):

First, we create the imaginary dataset, using the replicate function and a random number generator:

```
TestData<-tibble(
  DayTime=now()+minutes(rep(c(0:359),100)),
  Obs=rnorm(36000))
```

We can calculate the hourly average by piping our `TestData` tibble (above) through a group_by and then a `summarise()` function.

```
TestData %>%
  group_by(yday(DayTime),hour(DayTime)) %>%
  summarise(Mean=mean(Obs))
```

`summarise()` has grouped output by 'yday(DayTime)'. You can override using the `.groups` argument.

```
# A tibble: 7 x 3
# Groups:   yday(DayTime) [1]
  `yday(DayTime)` `hour(DayTime)`     Mean
            <dbl>           <int>    <dbl>
1             360              13 -0.0953
2             360              14  0.00116
3             360              15 -0.00792
4             360              16  0.00323
5             360              17 -0.0208
6             360              18 -0.0237
7             360              19 -0.00627
```

Question Why do we include yday() in the group_by function?

Answer: Remove yday(DayTime) and compare the output.

Chapter 9

R Markdown

9.1 Overview

By now you have mastered the fundamentals of base R, visualizations, and data science!

R Markdown is a powerful format for quickly making high-quality reports of your analysis. You can embed code and all kinds of output, including graphs, and output them to a Word Document, PDF or website. In fact, this entire book was written in *R Markdown*!

Here we'll cover just the basics, but a complete guide to R Markdown is available online from *Yihui Xie, J. J. Allaire and Garrett Grolemund* (*https://bookdown.org/yihui/rmarkdown/*). You can also check out the R Markdown documents that we use to make our tutorial websites on our GitHub Pages (the website files have .html extension and the R Markdown files have the same name with .Rmd extensions):

- Colautti Lab Resources Website: *https://colauttilab.github.io/*
- Colautti Lab GitHub Repository: *https://github.com/ColauttiLab/ColauttiLab.github.io*

9.2 Setup

Before beginning this tutorial, make sure you have installed these packages:

```
install.packages('rmarkdown')
install.packages('dplyr')
install.packages('knitr')
```

These should be installed with R Studio, but you may want to re-install them if you are working with an older version. You may have to quit and restart R Studio a few times during this process.

9.3 Cheat Sheet

There is a very handy 2-page 'cheat sheet' (*https://www.rstudio.com/wp-content/uploads/2015/02/rmarkdown-cheatsheet.pdf*), which you can also access through R Studio under the *Help* menu: `Help -> Cheatsheets`. There are several other links here too, including `ggplot` and `dplyr`.

9.4 Create

So far, you've been working through the command line in a `.R` program file. As discussed in the *R Fundamentals* Chapter, the benefit of programming in a `.R` file is that we can share the program and associated files with other computers, including high-performance servers.

An R Markdown file is another special text file, with a `.Rmd` extension. It's useful for generating reports with embedded code and visualizations. I use it as a virtual notepad when analyzing new data, and later convert it to Online Supplementary Material for published papers. The utility of the `.Rmd` file is easiest to understand by example.

Because an R Markdown file is just a text file with a `.Rmd` extension, we could just make a new text file and save it with the `.Rmd` extension. However, in RStudio, we can make a new R markdown file from the menu: `File-> New-> R Markdown`

Choose `Document` from the left-hand side menu, give it a title, and make sure `html` is selected.

Then click `OK`

Very few elements are needed for a basic markdown file, and examples of these elements are provided within the R Markdown file that R Studio sets up for you.

At the top of this window is a little icon that says 'Knit'. The **Knit** icon, knits together your text file into a rendered html document.

Try clicking the *knit* icon to generate a report using the default text in the R Markdown file. Note that you may have to save a copy of your R Markdown file first. A new file will open in a web browser. Take a moment to compare the output file with the input R Markdown.

If you want to output this as a pdf file, then you can simply choose `Print`, and then `Print to PDF` in your web browser. You can also make pdf files directly from R Studio, but you might run into problems depending on your file content and format.

We'll stick with the html version for now, and walk through some of the main components available to you.

9.5 YAML Header

Every R Markdown file starts with a **YAML header**, which contains some basic information about the file. A YAML header is generated automatically when you make a new .Rmd file in RStudio, but not all elements are needed. Depending on what options you choose, it might look something like this:

```
---
title: "Untitled"
author: "Robert I. Colautti"
date: "January 20, 2019"
output: html_document
---
```

There are other options available for YAML, and you can includes a separate `_output.yml` to set other aspects of the layout.

Here are some additional formatting options. Replace the `output: html_document` line above:

```
output:
  html_document: # Add options for html output
    toc: true # Add table of contents (TOC)
    number_sections: true # Add section numbers
    toc_float: # Have TOC floating at the side
      collapsed: false # Expand subsections
```

9.6 Markdown Elements

The *Markdown* in R Markdown refers to the **Markdown protocol** (*https://en.wikipedia.org/wiki/Markdown*), which is a non-proprietary system that was designed to quickly and easily encode formatted documents and websites in a simple text document.

The main advantage of R Markdown (`.Rmd`) over regular Markdown (`.md`) is the ability to easily print, format, and execute embedded R code for graphs, tables, and calculations.

We'll look at some basic Markdown elements and then we'll see how to embed R code for professional, reproducible reports.

9.7 Basic elements:

9.7 Plain text

Plain text is converted into paragraph format.

To start a new paragraph, press *enter* twice. This is important – if you only press enter once, then the two paragraphs will knit together into the same paragraph.

Similarly, if you inlcude more than two lines between paragraphs, these will be ignored when you render the R Markdown document.

Try adding paragraphs of text separated by pressing enter 1, 2 and 3 times. Then, knit to html to see how these are rendered in the final output.

9.7 Formatted text

You can format text with * or _

`*italics*` or `_italics_`: *italics*

`**bold**` or `__bold__`: **bold**

Use greater-than sign for block quotes, eg. `> TIP: quote`

> TIP: quote

9.8 Headers

Add headers with up to six hash marks (#). Each additional # denotes a sub-heading of the previous (sub)heading.

`# Header 1`

`## Sub-Header = Header 2`

`### Sub-Sub Header = Header 3`

`#### Sub-Sub-Sub Header = Header 4`

9.9 Other Elements

Use two dashes (--) for short–dash (a.k.a. 'n-dash').

Use three dashes (---) for long — dash (a.k.a. m-dash).

9.10 Links

Links have a special format. The text you want the user to see goes in square brackets, followed immediately by the file or html link in regular brackets, with no space in between. You can use both web links and relative path links.

`[Colautti Lab Website](https://colauttilab.github.io/)`

This should produce a link if you are reading this electronically:

Colautti Lab Website

You can also use this with relative path names, for example to link a file in a folder called `images` inside of the project folder:

`[Linked .png file](./images/ColauttiLabLogo.png)`

Again, this should produce a link if you are reading this electronically. However, you will get an error if you try to include this in your R Markdown file.

> **Question**: Why will you get an error?

Answer: This file is not in your working directory. You will need to download this image and save it as `ColauttiLabLogo.png` inside a directory called `images`, which is also inside of the working directory that contains your `.Rmd` file.

Linked .png file

9.11 Images

Instead of linking, you can embed the image directly by adding an exclamation point. The text in square brackets becomes the figure caption.

`![Linked .png file](./images/ColauttiLabLogo.png)`:

Note that R Markdown added a figure number for me, based on the chapter and the number of previous images. This doesn't include the graphs that were created with embedded R code, only images that were embedded with `![]()`.

Figure 9.1: Linked .png file

9.12 Lists

Lists are easy to create, simply start a line with * or + for *unordered* lists or a number for *ordered* lists. Add tab characters for sub-lists:

```
+ Unordered list item 1
* Item 2
    + sub item 2.1
    * sub item 2.2
* Item 3
```

- Unordered list item 1
- Item 2
    - sub item 2.1
    - sub item 2.2
- Item 3

```
1. Ordered list item 1
2. Item 2
    + sub item 2.1
    * sub item 2.2
3. Item 3
```

9.12. LISTS

1. Ordered list item 1

2. Item 2

   - sub item 2.1

   - sub item 2.2

3. Item 3

The fun thing about ordered lists is the numbers you use don't really matter – R Markdown will automatically start at 1 and increase for each item.

```
1. Ordered list item 1
1. Item 2
   + sub item 2.1
   * sub item 2.2
1. Item 3
```

1. Ordered list item 1

2. Item 2

   - sub item 2.1

   - sub item 2.2

1. Item 3

This is a nice feature in early drafts of your R Markdown, to which you might later add, rename, or reorganize the order. You can just leave the numbers and not waste effort renumber each time.

9.13 Tables

Tables are added using a line of horizontal dashes to separate the title of the table, then a row of header names separated by pipes to define the header row. Finally, we add another line of dashes with pipes to indicate the relative column widths. The data goes underneath, with pipes separating each column.

```
Tables
-----------------------
Date     | Length | Width
------|---------|------
09/09/09 | 14 | 27
10/09/09 | 15 | 29
11/09/09 | 16 | 31
```

Produces this output:

9.14 Tables

Date	Length	Width
09/09/09	14	27
10/09/09	15	29

Date	Length	Width
11/09/09	16	31

Now that we have written some basic markdown elements, we can generate our R markdown report to see what they look like. Click on the *knit* button and compare the input with what you have typed in your R markdown file.

9.15 Embed R Code

You can format text to look like code using the back-tick character.

```
# `Use single tick mark to invoke code formatted text.`
```

The **Back Tick** is a strange looking character usually located on the same key as the tilde (~) on English keyboards. Don't confuse the back tick with the single quotation mark.

You can incorporate blocks of R code using three *back ticks* with *r* in curly brackets. Write some code, then add three more tick marks to signal the end of the code chunk. By defaule, your code will run when you convert your R Markdown to html, showing the output. This is a great way to include graphs and the output of statistical models.

```
#    ```{r}
#    <your code goes here>>
#    ```
```

`Ctl-Alt-i` is a nice shortcut in R Studio for adding code chunks quickly.

9.15 Code Chunk Names

You can name your code chunks by adding a name right after the r separated only by a space. Naming code chunks is very handy for troubleshooting. When you *knit* your file, the name of each chunk is listed in the *Render* tab in R Studio. If there is an error, you can see which code chunk is causing the error. Note that the name cannot contain spaces, and it can be followed by a comma to specify options for the code chunk.

```
# ```{r code-chunk-name, eval=F}`
```

9.15 Code chunk parameters

You can use different options for your R code chunks, as shown on the R Markdown cheat sheet. The default is to both print the code AND show the output, but these can be changed with options. The three options I use most commonly are:

- eval=F – Don't *evaluate* the code. The code will be shown, but it won't be run so no objects or output will be created.

- echo=F – Don't *echo* the code. Don't show the code, but run it and include any output, plots, messages and warnings.

- include=F – Don't *include* the code in the rendered document. Run the code to create any objects and load libraries, but don't show the code or any output, plots, messages or warnings.

9.16 Dynamic tables

Making tables from data is a bit more complicated. For example, if we wanted to summarize the `FallopiaData.csv` data, we could read in the file and then summarize with dplyr as we did in the *Data Science* Chapter.

```
library(dplyr)

Fallo<-read.csv(
  "https://colauttilab.github.io/RCrashCourse/FallopiaData.csv")

SumTable<-Fallo %>%
  group_by(Taxon,Scenario,Nutrients) %>%
  summarize(Mean=mean(Total), SD=sd(Total)) %>%
  arrange(desc(Mean))

print(SumTable)
```

```
# A tibble: 10 x 5
# Groups:   Taxon, Scenario [10]
   Taxon Scenario     Nutrients  Mean    SD
   <chr> <chr>        <chr>      <dbl> <dbl>
 1 bohem high         high       60.3   8.68
 2 japon gradual      high       59.7   9.57
 3 bohem fluctuations high       58.4   9.20
 4 bohem extreme      high       58.3   7.34
 5 bohem gradual      high       57.5   9.34
 6 japon extreme      high       57.2  10.9
 7 japon high         high       56.4   8.20
 8 japon fluctuations high       56.4  13.7
 9 japon low          low        52.0   8.29
10 bohem low          low        48.0   8.86
```

Table 9.2: Summary Table

Taxon	Scenario	Nutrients	Mean	SD
bohem	high	high	60.28091	8.677075
japon	gradual	high	59.72917	9.565376
bohem	fluctuations	high	58.36455	9.202334
bohem	extreme	high	58.30917	7.337015
bohem	gradual	high	57.46154	9.338311
japon	extreme	high	57.23643	10.903133
japon	high	high	56.44833	8.204091
japon	fluctuations	high	56.43692	13.724906
japon	low	low	52.02917	8.287938
bohem	low	low	47.98077	8.862164

The output is legible but not very attractive for a final report. To make it look better, we can use the `kable` function from the `knitr` package:

```
library(knitr)
kable(SumTable, caption = "Summary Table")
```

The output of this code is shown in the Summary Table. Contrast the formatting from the `kable()` function with the standard R output from `print()`.

9.17 Embed Graphs

Use R code to embed graphs.

```
ggplot() +
  geom_histogram(aes(rnorm(100)), binwidth=0.2)
```

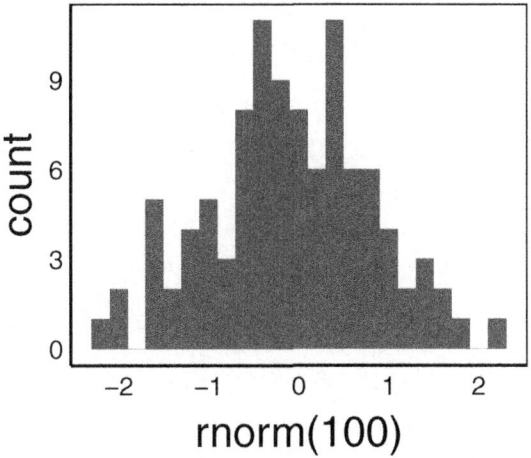

9.18 Content as tabs

Rendering an html has additional features that won't show up in a pdf. The code below will produce tabs in your knit html file.

```
## Tabs by Group {.tabset}

### Treatment Group

1. Treatment 1 explanation
2. Treatment 2 explanation
3. Treatment 3 explanation

### Histogram

#```{r}
#library(ggplot2)
#ggplot() +
#   geom_histogram(aes(rnorm(100)),binwidth=0.2)
#```
```

The output of this code is shown in the figure.

Tabs by Group

Treatment Group | Plot 1

```
library(ggplot2)
ggplot() +
  geom_histogram(aes(rnorm(100)),binwidth=0.2)
```

Figure 9.2: Tabs created in html using {.tabset} with ## headers. The "Plot 1" tab contains embedded R code to produce a histogram graph. Both the code and graph are shown by default.

9.19 Equations

Insert equations using LaTeX equations. LaTeX is a system for converting basic text files to formated documents.

Use single dollar signs for in-line equations, like $Y = X$, which will print as $Y = X$ on the same line as the text.

2. Use double dollar signs on a new line for full-line equations, like $$Y = X$$:

$$Y = X$$

which isolates the equation on its own line, and centers it.

You can use a variety of Greek letters by using the backslash character \. For upper-case Greek letters, just use an upper-case letter in the spelling. For example:

- \lambda(λ) OR \Lambda (Λ)
- \gamma(γ) OR \Gamma(Γ)
- \delta(δ) OR \Delta(Δ)

Omicron can be \omicron or simply the letter o, with no backslash o (o). Note that some LaTeX characters, like omicron, do not have capital versions (e.g., \Omicron does not produce a Greek character).

There are numerous other options but below is a quick rundown of some of the commonly used scripts.

Script	Description	Code	Example
\infty	Infinity	\infty	∞
_	Subscript	X_i	X_i
^	Superscript	X^2	X^2
'	First order derivative	f'(x)	$f'(x)$
''	Second order derivative	s'(x)	$s'(x)$
\sim	Predict	Y \sim X	$Y \sim X$
\times	Multiply	X \times Y	$X \times Y$
\pm	Plus or minus	X \pm Y	$X \pm Y$
\neq	Note equal	X \neq Y	$X \neq Y$
\leq	Less than or equal	X \leq Y	$X \leq Y$
\geq	Greater than or equal	X \geq Y	$X \geq Y$
{}	Group together	X_{subscript}	$X_{subscript}$
\sqrt	Square root	\sqrt{x^2y^2}	$\sqrt{x^2y^2}$
\frac	Fraction	\frac{X+1}{X-1}	$\frac{X+1}{X-1}$
\sum	Sum	\sum_{x=1}^{K}	$\sum_{x=1}^{K}$
\prod	Product	\prod_{x=1}^{K}	$\prod_{x=1}^{K}$
\int	Integral	\int_{0}^{\infty}	$\int_{0}^{\infty}$
\lim	Limit	\lim_{x \to \infty}	$\lim_{x \to \infty}$

Note in particular, the use of curly brackets to group items together in superscripts, subscripts, fractions and square root. Also note the *simulate* (\sim) character, which is the *tilde* (~) used in statistical models and other R functions like facet_grid() and aggregate().

Here are some more sophisticated examples to show how to create more complex equations. Again, try reproducing these in R markdown. If you don't get the same output, then check to see what is different with your code.

9.19. EQUATIONS

`$$Y_i \sim \alpha + \beta_1 X_i + \epsilon_i$$`

will produce a linear model equation:

$$Y_i \sim \alpha + \beta_1 X_i + \epsilon_i$$

and

`$$sum_{n=1}^{\infty} 2^{-n} = 1$$`

will produce:

$$\sum_{n=1}^{\infty} 2^{-n} = 1$$

Note the use of special characters with the backslash \, along with subscripts _ and superscripts ^ with text in curly brackets {}.

That's all you need to know to produce professional reports with R and R Markdown!

Chapter 10

Custom Functions

10.1 Overview

So far, everything you have done in this book applies functions that somebody wrote. From basic `c()` to advanced functions in the packages `dplyr`, `ggplot` and `lubridate`, all of these were written by somebody. Now it is your turn.

Custom functions are useful whenever you find yourself repeating code. For example, maybe you are repeating a calculation across different data sets. Or maybe you are repeating code with slight changes to some of the parameters each time. Condensing repeated code into functions can help make your code more concise, organized, and understandable. Functions can also make repeated code run faster, as we very briefly introduced in the *Flow Control* chapter. Functions can also help to avoid errors, because you only change parameters in one place – when you run the code – instead of replacing parameters in all of the steps of the function.

Giving your custom function a clear name and specifying the arguments that it takes, will help make your code easier to understand. This is important for collaboration, even if you are collaborating with yourself at some time in the future.

Custom functions can be a little tricky to master at first, but you've already taken the biggest step by learning how to run functions that others have written. By digging into the help files of the functions you know, you should get a good feel for what functions can do and how they are organized. In this chapter, you will learn how to build your own functions, and some of the ways they may be particularly useful to you.

10.2 General form:

You should already have a good sense of how functions work from all of the other tutorials/chapters. Now let's work through a real example. Don't type this out, but read through it:

```
functionName<-function(var1=Default1,var2=Default2){
  ## Meat and potatoes script
  return(output)
}
```

This is called pseudo-code. The purpose is to give you a general sense of how to create functions. We have variables (`var1` and `var2`) and we can assign default values (`Default1` and `Default2`). The comment `## Meat and potatoes script` represents the main steps of the function. The final `return(output)` contains the output that is *returned* to the user of the function.

The different components of the custom function are easier to understand by example.

10.3 Example function

We'll convert the code into a real example, where the user will input two numeric objects. The function outputs a list of functions applied to the inputs.

```r
my.function<-function(var1=0,var2=0){
  # We can make new variables within a function
  add<-var1+var2
  subt<-var1-var2
  mult<-var1*var2
  div<-var1/var2
  # And put them together into a list for output
  outlist<-list(input1=var1, input2=var2,
                addition=add, subtraction=subt,
                multiplication=mult, division=div)
  # So far, everything is contained within the function.
  # Use return() to generate output
  return(outlist)
}
```

On the first line, we define the function by giving it a name (my.function) and setting parameters for the function. In this function we have just two parameters, one for each input variables. If one or both are left blank when we call the function, then R will replace these values with the default (zero). These variables may be individual numbers, or we may input vectors and R will automatically apply them to each element.

Inside of the function, we generate five objects, the first four representing simple mathematical equations applied to the two input variables. A fifth object (outlist) is simply a list object containing the two input variables and the output of each of the four equation objects.

The final line `return()` contains the object that is output from running the custom function. In this case, it is the `outlist` list object.

As usual, be sure to type out this function in a `.R` script or a code chunk in a `.Rmd` file. Try running the first line of the function. You'll see a + sign in your R console. This is R telling you that it is expecting more lines of code. This happens when you have open brackets or an unfinished pipe (`%>%`) or ggplot (`+`) command. Run each of the remaining lines of the function, and you should see the R Console return to > after you run the last line of the function.

Question: Why is there no output to the R Console?

Answer: You have just loaded the function into memory. Think of this like when you use `library()` to load a package into memory. The functions from that package are now available for use.

Look at the *Environment* tab in R Studio. This is a tab in one of the R Studio windows – usually in the top-right window by default. You should see a new item here called `my.function` followed by `function (var1 = 0, var2 = 0)`. This tells us that the `my.function()` function is available for use, and it has two input parameters with default values.

10.4 Local vs Global

Note what is missing from the *Environment* tab: none of the objects that are created inside the function are listed here. For example, `add`, `subt` and `outlist`. Even when we run the function, we won't see those objects in the Environment here. These objects are **local objects** because they only exist within the function that contains them.

10.5. RUN CUSTOM FUNCTIONS

By contrast, a **global object** is created when we make an object in the main code.

```
Global<-"Object"
```

Global objects are saved in memory and can be accessed by any function that you run. **Local objects** can only be used inside of the function that contains them.

10.5 Run custom functions

Running custom functions is no different from running any of the other functions you are familiar with. Try running the function on its own, with default values:

```
my.function()
```

```
$input1
[1] 0

$input2
[1] 0

$addition
[1] 0

$subtraction
[1] 0

$multiplication
[1] 0

$division
[1] NaN
```

Now try specifying the input parameters and compare the output.

`my.function(var1=10,var2=0.1)`

```
$input1
[1] 10

$input2
[1] 0.1

$addition
[1] 10.1

$subtraction
[1] 9.9

$multiplication
[1] 1

$division
[1] 100
```

`my.function(var1=c(1:10),var2=c(10:1))`

```
$input1
 [1]  1  2  3  4  5  6  7  8  9 10

$input2
 [1] 10  9  8  7  6  5  4  3  2  1

$addition
 [1] 11 11 11 11 11 11 11 11 11 11

$subtraction
 [1] -9 -7 -5 -3 -1  1  3  5  7  9
```

```
$multiplication
[1] 10 18 24 28 30 30 28 24 18 10

$division
[1]  0.1000000  0.2222222  0.3750000  0.5714286  0.8333333
[6]  1.2000000  1.7500000  2.6666667  4.5000000 10.0000000
```

10.6 Annotation

For more complicated functions that take a long time to run, consider using `print()` or `cat()` to indicate the steps that are being run. This can help a lot with troubleshooting custom functions. The `cat` function is similar to `print` but lets you print directly to screen rather than passing through a data object. Recall from the *Regular Expression* Chapter that \n is the *new line* character. If we include \n in the `cat()` output, then it will print to a new line. Here is an example:

```
my.function<-function(var1=0,var2=0){
  cat("\nInput variables:\nvar1 =", var1,"\nvar2 =", var2,"\n")
  cat("\nCalculating functions...\n")
  cat("\nAdding...\n")

  add<-var1+var2

  cat("\nSubtracting...\n")

  subt<-var1-var2

  cat("\nMultiplying...\n")

  mult<-var1*var2
```

```
    cat("\nDividing...\n")

    div<-var1/var2

    cat("\nGenerating output...\n\n")

    outlist<-list(input1=var1, input2=var2,
              addition=add, subtraction=subt,
              multiplication=mult, division=div)

    return(outlist)
}

## Run
my.function(var1=10,var2=0.1)
```

```
Input variables:
var1 = 10
var2 = 0.1

Calculating functions...

Adding...

Subtracting...

Multiplying...

Dividing...

Generating output...

$input1
[1] 10
```

```
$input2
[1] 0.1

$addition
[1] 10.1

$subtraction
[1] 9.9

$multiplication
[1] 1

$division
[1] 100
```

10.7 Verbose parameter

Printing text to the screen can slow down your function considerably, as we saw in the *Flow Control* Chapter. A good practice is to provide output as a user-defined option by adding a 'verbose' parameter and an if() statement.

```
my.function<-function(var1=0,var2=0,verbose=FALSE){
  if(verbose==T){
    cat("\nInput variables:\nvar1 =", var1,"\nvar2 =", var2,"\n")
    cat("\nCalculating functions...\n")
    cat("\nAdding...\n")
  }

  add<-var1+var2

  if(verbose==T){
    cat("\nSubtracting...\n")
```

```
    }

    subt<-var1-var2

    if(verbose==T){
       cat("\nMultiplying...\n")
    }

    mult<-var1*var2

    if(verbose==T){
       cat("\nDividing...\n")
    }

    div<-var1/var2

    if(verbose==T){
       cat("\nGenerating output...\n")
    }

    outlist<-list(input1=var1, input2=var2,
                  addition=add, subtraction=subt,
                  multiplication=mult, division=div)

    return(outlist)
}
```

Now the Outlist is returned, but the `cat()` functions are only run if Verbose=T is selected when running the function.

10.8 External files

In the *Basic Customization* Chapter, we saw how to create a custom plotting theme and save it as a file that we could load to apply the theme.

10.8. EXTERNAL FILES

The same is true for custom functions.

If you have a custom function that you would like to use frequently, or if it is too big to include in your main R Script or R Markdown file, then a dedicated .R file may be a good option.

1. **Save** in a separate file, typically with a .R extension. For example, we might make new R Script called `myfunction.R` containing just the lines of the `my.function()` function that we created earlier.

2. **Load** using `source("PathName.FileName.R")`. For example, we may have a directory called `scripts` inside of our working directory, in which case we could load the custom function with `source("./scripts/myfunction.R")`.

Chapter 11

Conclusion

11.1 You made it!

Congratulations, you made it to the end of the book! Now you know all of the R fundamentals that I wish I had learned when I was an undergraduate working on my first research project. You may not feel like an 'expert' yet. However, take a moment to reflect on everything that you've accomplished by working through this book. Seriously, just sit down with a pen or pencil and paper and try to make a list of everything you've learned. This is an important self-reflection exercise.

Okay, now I'll try. First, you learned how to program in the command line by typing commands into the R Console in R Studio to produce an output. That alone is a major hurdle for many biologists! But you went even further, perhaps confronting hesitation or self-doubt about your ability to work with mathematical equations. You've learned how to translate equations into code, from simple addition, subtraction and multiplication to specific functions for the absolute value, square root, log, average, variance, and others. More importantly, you've learned the principles of coding that allow you to translate just about any equation into your very own custom R function. You got tied up with brackets, and brackets within brackets, within brackets to code more complicated equations.

You learned how to use logic operators with flow control and `dplyr` commands to string your functions together into an automated and reproducible workflow. You even learned how to write your own custom functions!

You learned all about data frames (i.e. *flat* data or *2-dimensional* data) and all the useful functions from the `dplyr` library, like sub-setting, joining, sorting, grouping, and summarizing. You may not realize it, but this taught you how to work with *relational data*. Maybe you don't call it this, but relational data are just data that *relate* to each other, which you learned to do with `dplyr` and `join_`. This is not much different than what advanced coders and data scientists do with large database protocols like *Hadoop* or *SQL*. These lie at the heart of user-friendly web tools for interacting with large online databases you may be familiar with, like *Climate Data Online (CDO)*, the *Global Biodiversity Information Facility (GBIF)*, or the many genetic and genomic databases maintained by the European *Molecular Biology Laboratory (EMBL)* and the *National Center for Biotechnology Information (NCBI)*.

You learned how to work with naughty data, like missing values, miscoded entries, and dates. Oh man, dates probably caused you so many problems until you leaned how to deal with them in R. You will never, ever, ever, work with dates in *Microsoft Excel*, if you can help it!

You ventured into the intimidating and enchanting world of regular expressions. Once you start looking for opportunities to use regular expressions, you will start to see them in just about every data project you have, whether its pulling out data with specific characteristics, or reformatting database entries by those who don't understand the importance of strict data encoding practices. It's unlikely you've masted

11.1. YOU MADE IT!

regular expressions, especially if this was the first time you were exposed to these powerful spells. The ability to code regular expressions will come with practice, which will enhance your power to bring good to the world, much like the *Expecto Patronum*.

You learned everything you need to know to produce professional, publication-worthy visualizations of your data. You learned about important graphical concepts like data formats and colour palettes for publishing, and the layered 'grammar of graphics' philosophy. Even better, you learned how to wrap it all together in a reproducible and professional report with *R Markdown*.

Above all, you leaned how to embrace mistakes and troubleshoot your coding problems.

Although you have learned a lot, it's natural to feel confused or unclear about some of the concepts and techniques you have encountered. In many cases, you may not yet be aware of your knowledge gaps.

That's okay! You are now officially a real coder. Research is a lifelong learning process, and you will continue to encounter new challenges and opportunities to learn new tricks and techniques as you progress along your coding journey. Sure, you have much to learn, but so does everybody else. Maybe you'll meet a coder with a degree in computer science and a decade of experience coding for a major tech corporation. They will understand coding better than you, but you'll probably be able to discuss coding better than they can discuss biology. Don't get intimidated. Embrace the journey, not the destination.

In short, you didn't just learn how to code, you learned *how to learn* how to code. And, you deserve a massive, giant congratulations!

11.2 What next?

Where do you go from here?

First, treat yourself. Maybe take a vacation? You deserve it.

Then, remember that you are now a competent coder. You have a lot to learn, but you don't have to wait until you feel like an 'expert' to put your skills to good use. Look for opportunities to write and proofread code to hone your skills. Personally, I have found that offering my time to teach others, to help others with their troubleshooting, and to collaborate on projects, have all made me a much better coder, and a well-rounded researcher.

If you found this book suited your learning style, then you might want to check out some of the other books and resources that we are developing.

1. `EcoEvoGeno.org` is our main, public-facing website and it contains links to our latest book releases as well as general information about our research and lab activities.

2. `ColauttiLab.github.io` is our lab 'resources' website, with a variety of links for new and established coders. It's directed at the graduate students in our lab, but many of these resources may be useful to you.

3. `github.com/ColauttiLab` is our github page, where you can search our repositories to see what we are currently working on.

4. *R STATS Crash Course for Biologists* is the next book in this series. If you are going to work with biological data in R, then you probably will want to learn how to run statistical models. The *R STATS*

11.2. WHAT NEXT?

Crash Course for Biologists covers this, from the most basic ANOVA or linear regression all the way to cutting-edge Generalized Additive Mixed Effects Models.

5. The third book in this series is *R Machine Learning Crash Course for Biologists*. Once you understand statistical models, the *R Machine Learning Crash Course for Biologists* will guide you through common *supervised* and *unsupervised* machine learning models, including a deep-dive into the Principal Components Analysis (PCA) mentioned only briefly in the *R Fundamentals* Chapter. In addition, you'll learn how to run Regularized Discriminant Analysis, Support Vector Machines, and Decision Trees to make predictions.

6. The *Python Crash Course for Biologists* is pretty much the same as this book, but written in Python instead of R. Once you are comfortable coding in R, then you will see it is actually very easy to move into Python.

If you have caught the coding bug and you would like to dive deeper into pure R coding, then I would suggest anything by Hadley Wickham. If you have ideas for additional content that you would like to learn, but you can't find a good resource, then please don't hesitate to reach out.

If you have any thoughts you would like to share, good or bad, please get in touch. If you have criticisms, please send them to me so that we can improve future editions of the book. You can find up-to-date contact information on our lab website.

11.3 Support Open & Accessible Science

If found this book helpful, please consider supporting us. We have avoided using big name publishers so that we are able to share our content at no charge, as we do on Github and in training sessions – both in-person and virtual. We have also tried to keep the cost low for our printed versions and the proprietary electronic versions (e.g. Kindle, Kobo, Apple, Google). Rather than pay publishing cartels or professional editors, all of the proceeds from these versions support graduate students to help teach and develop new content, including translations to other languages and beta testing new tutorials for future books.

Our team is passionate about demystifying math and coding for biologists, and we want to make these skills more accessible to the next generation of biologists, empowering students of all backgrounds and historically under-represented groups in particular.

If you would like to support us, please consider buying a copy to gift to a friend or colleague, if it is within your means. If your budget is tight, then please consider posting a thoughtful and supportive review on Amazon, Barnes & Noble, Apple Books, Google Play Books, or wherever you read this. A positive review will help others to find the book, which will help to build our small community of biology coders. If you aren't comfortable posting a 5-star review, please contact us to let us know what we can do to bring the next iteration of this book up to your standards.

As you develop your coding skills, consider making recommendations to help improve our books. The most efficient way to do this is by posting an *issue* in Github. Alternatively, you can find up-to-date contact information on our lab website.

As you continue on your journey, remember that learning to code is different from most biology that you've learned. To *really* learn to code, you must continue to immerse yourself, study, read, try something new, fail, correct, and repeat. And of course: practice, Practice PRACTICE!

Thinking back on what I've learned in coding in R, Python, and Unix since 2009, one thing sticks out as particularly helpful for solidifying my understanding of code: helping others. This book began as a series of self tutorials to teach coding to biologists. This came on the heels of helping with full-day coding workshops with what is now called the *Centre for Advanced Computing* at Queen's University. It continues all the way back to my experience as a graduate TA helping with statistics, and offering help to other graduate students who were new to R. All of these experiences helped to reveal blind spots in my learning and offered opportunities to practice my skills. If you found this book helpful, and you want to continue to develop your skills, this is the best advice I can offer: **Pay it forward**.

11.4 Picture a Coder

Finally, please share your knowledge and experiences with others. As you continue to learn and explore R programming, consider sharing your insights and discoveries with your peers and colleagues. This will help to build a supportive community, and you will probably find that helping others helps you hone your own skills. You never know who you might inspire to embark on their own coding journey.

Let us conclude by reviewing your answer to the preface of this book when you were asked to Think of a computer programmer or data scientist.

Question: What does a computer programmer look like?

Can you picture yourself in that role? If you completed this book, you should!

Printed in Great Britain
by Amazon